AF352841

Effects of Mineral Elements on the Environment

Effects of Mineral Elements on the Environment

Editors

Marina Cabral Pinto
Amit Kumar

MDPI • Basel • Beijing • Wuhan • Barcelona • Belgrade • Manchester • Tokyo • Cluj • Tianjin

Editors
Marina Cabral Pinto
Department of Geosciences
University of Aveiro
Aveiro
Portugal

Amit Kumar
School of Hydrology and Water
Resources
Nanjing University of
Information Science &
Technology
Jiangsu
China

Editorial Office
MDPI
St. Alban-Anlage 66
4052 Basel, Switzerland

This is a reprint of articles from the Special Issue published online in the open access journal *Applied Sciences* (ISSN 2076-3417) (available at: www.mdpi.com/journal/applsci/special_issues/mineral_elements_environment).

For citation purposes, cite each article independently as indicated on the article page online and as indicated below:

LastName, A.A.; LastName, B.B.; LastName, C.C. Article Title. *Journal Name* **Year**, *Volume Number*, Page Range.

ISBN 978-3-0365-1788-9 (Hbk)
ISBN 978-3-0365-1787-2 (PDF)

Contents

About the Editors

Marina Cabral Pinto

Dr Marina Cabral-Pinto works in environmental geochemistry research on an array of topics such as medical geology, epidemiology, water, soil, air quality, geo-bioremediation, ecological health assessment, carbon sequestration, climate changes, etc. She completed her degree in Geological Engineering, MsD (University of Aveito and University of Coimbra), and obtained her Ph.D and Postdoc from University of Aveiro, where she is now working as an investigator and invited professor. During her academic career, she has published more than 60 research/review articles of international repute and attended several conferences and training events . In addition to this, she has handled Special Issues of *IJPREH* (MDPI), *Land* (MDPI), *Applied Sciences* (MDPI), and *Mine Water Association* (Springer). She has supervised more than 15 PhD and MSc students. She was granted with the Orlando Leitão premium by the Portuguese Society of Neurology due her research and results on links between environmental potentially toxic elements and cognitive disorders and diseases, such Alzheimer's disease. She was funded by FCT for the Covid19 Project and by CNRS for the LabEx DRIIHM research: Exposure and Health in an industrial area (Portugal); ChildIngest –Ingestion of dust by young children: physicochemical characterization, bioaccessibility and genotoxicity: https://orcid.org/0000-0002-6908-1596
Scopus Author ID: 22133337700
Ciência ID: B110-D819-0C69
https://www.researchgate.net/profile/Marina-Cabral-Pinto/publications?sorting=recentlyAdded &editMode=1Delivery.

Amit Kumar

Dr. Amit Kumar (Associate Professor) is an environmentalist-cum-hydrologist who has worked on wide areas such as sources, fate and transport of carbon in water and soil, greenhouse gas emissions from reservoirs/ lakes/ forest, water quality and ecological health assessments, etc. He completed his M.Tech and Ph.D from one of the top technical institutes of India, IIT Roorkee, under the umbrella of the MHRD fellowship, and undertook postdoctoral research at Hohai University (College of Hydrology and Water Resources, Nanjing, China) and is now working with Nanjing University of Information Science and Technology, School of Hydrology and Water Resources, Nanjing, China, as an associate professor. During his academic career, he has published more than 65 research/review articles of international repute and attended several training events and conferences with the fellowship of ICAR, ETH, CAS, MOEF, DAAD, etc. In addition to this, he has handled Special Issues in *Land* (MDPI), *Applied Sciences* (MDPI), and *Frontier in Environmental Science,* and holds an honorary post as a Topic Editor for *Water* (MDPI) and *Sustainability* (MDPI), is an advisory board member of *Ecological Indicator*, Elsevier, and an associate editor of *Applied Water Science* (Springer) and *IJRBM* (Taylor and Francis).

Article

Prediction of Long-Term Health Risk from Radiocesium Deposited on Ground with Consideration of Land-Surface Properties

Hiroshi Yasuda 

Research Institute for Radiation Biology and Medicine, Hiroshima University, 1 Kasumi 2-3, Minami-ku, Hiroshima 734-8553, Japan; hyasuda@hiroshima-u.ac.jp

Featured Application: More appropriate decision making on how to manage a land contaminated by hazardous materials would be possible by application of the site-specific approach for health risk assessment presented herewith.

Abstract: After the Fukushima Daiichi accident, there have been long controversial discussions on "how safe is safe?" between the authorities and the residents in the affected area. This controversy was partly attributable to the way the authorities made a judgement based on the annual effective dose rate; meanwhile, many of the local residents have serious concerns about future consequences for their health caused by chronic radiation exposure, particularly of small children. To settle this controversy, the author presents an approach based on long-term cancer risk projections of female infants, i.e., the most radiosensitive group, following land contamination by radiocesium deposition into ground with different surface conditions; the land was classified into three categories on the basis of decaying patterns of radiation dose rate: "Fast", "Middle", and "Slow". From the results of analyses with an initial dose rate of 20 mGy per year, it was predicted that the integrated lifetime attributable risk (LAR) of cancer mortality of a female person ranged by a factor of 2 from 1.8% (for the Fast area) to 3.6% (for the Slow area) that were clearly higher than the nominal risk values derived from effective dose estimates with median values of environmental model parameters. These findings suggest that accurate site-specific information on the behavioral characteristics of radionuclides in the terrestrial environment are critically important for adequate decision making for protecting people when there is an event accompanied by large-scale radioactive contamination.

Keywords: terrestrial environment; radiocesium; cesium-137; radioactive contamination; cancer risk; radionuclide; nuclear accident; nominal risk; decision making

check for **updates**

Citation: Yasuda, H. Prediction of Long-Term Health Risk from Radiocesium Deposited on Ground with Consideration of Land-Surface Properties. *Appl. Sci.* **2021**, *11*, 4424. https://doi.org/10.3390/app11104424

Academic Editors: Amit Kumar and Marina Cabral Pinto

Received: 4 April 2021
Accepted: 11 May 2021
Published: 13 May 2021

Publisher's Note: MDPI stays neutral with regard to jurisdictional claims in published maps and institutional affiliations.

1. Introduction

After the severe nuclear accident that occurred at the Fukushima Daiichi nuclear power station on 11 March 2011, more than 160,000 people from within a 20 km zone and the area having relatively high radiation levels were forced to evacuate and, as of March 2021, more than 40,000 people are still recognized as evacuees [1]. While it is certain that those evacuation measures prevented some excessive radiation exposure of residents [2–4], it has caused severe psychological distress in evacuees who were forced to stay outside their home towns/villages for years [5,6].

Currently, most of the evacuation orders have been lifted in the affected areas where the levels of radioactive contamination are quite low. However, it seems many residents have not returned to their previous homes partly due to persistent concerns about the possible adverse health effects of chronic radiation exposure, despite the fact authorities have tried to convince them that they are overly concerned about the radiological effects by introducing the recommendations of international authorities such as the International Commission on Radiological Protection (ICRP). This standstill is attributable to the way authorities have provided explanations by the using short-term effective dose estimates. For example, soon after the accident occurred, the national and local governments announced that the reference level for such an emergency situation was typically in the effective dose band of 20 to 100 mSv, with citation of the basic recommendations of ICRP [7]. Though updates of these reference levels have been recommended by ICRP in a recent publication [8] as shown in Table 1, the decision process based on effective dose of a certain-scale population has not essentially changed.

Table 1. Updated reference levels for guiding the optimization of protection of members of the public during the successive phases of a nuclear accident [8].

Early and Intermediate Phase	Long-Term Phase
100 mSv or below for the entire duration of both the early and intermediate phases [1]	Lower half of the 1 to 20 mSv per year band with the objective to progressively reduce exposure to levels towards the lower end of the band or below, if possible

[1] Previously, the Commission recommended the selection of reference levels in the band of 20–100 mSv for emergency exposure situations. The current recommendation recognizes that in some circumstances, the most appropriate reference level may be below 20 mSv.

With their own experience of risk communication with the people affected by the nuclear accident in Fukushima, the author feels that many of the residents have been suspicious of such uniformly applied, dose-based approach since they know well that every individual could have a unique consequence on his/her health from the same dose radiation and, in particular, one with a higher radiosensitivity, like a small child, could suffer from more severe damage. Actually, it has been clearly shown by many experts that the radiosensitivity of humans significantly changes depending on age and sex. In general, smaller children are more radiosensitive than adults, which is attributable to the fact that children have more actively dividing cells that could amplify radiation-induced DNA damage to a large extent compared to the adults [9–13]. It is also known that female people are more radiosensitive in regard to carcinogenesis, mainly because the breast is notably susceptible to radiation compared to other organs [7,9,11,13]. Thus, it is desirable to answer more sincerely to their underlying question, "how safe is safe?", by precisely showing the health risk for the most sensitive group on a community basis, instead of just presenting an average value of effective dose for an anonymous general population. It is also desirable to show the integrated risk of all hazards, including possible effects of other toxic materials such as heavy metals and some organic compounds released into the environment, and then to enable the authorities to make more appropriate decisions through comprehensive optimization of those different-quality risks and benefits.

With these considerations, the author tries to present an approach for predicting long-term radiological risk of female infants, i.e., the most radiosensitive group, under the assumption that they would live on the land contaminated by radiocesium (^{137}Cs) with consideration of the difference in land-surface properties. The radiological impact of ^{137}Cs, one of the most common products generated by the nuclear fission reaction, has been a subject of high concern in both Chernobyl and Fukushima Daiichi accidents because of its long half-life of about 30 years, chemically reactive feature, and high-energy (0.662 MeV) γ-ray emissions [14–16]. Meanwhile, other long-lived radionuclides such as ^{90}Sr (half-life: 29 years) and ^{106}Ru (368 days) could also be causes for concern depending on the situation of radioactive release to the environment, as indicated in the Chernobyl accident [17].

2. Materials and Methods

2.1. Projection of Absorbed Dose

First, a situation of land contamination caused by fall-out ^{137}Cs deposited onto the ground is assumed. In this situation, external exposure to γ-rays from the ^{137}Cs source in the ground surface is the main concern; internal exposure from radiocesium is supposed to be quite small as confirmed from the Fukushima Daiichi accident [3,4,18]; Kim et al. [18]

reported that most of the committed effective doses from radiocesium (^{137}Cs and ^{134}Cs) determined based on whole-body-counter measurements were less than 0.1 mSv for most of the examined residents (125 adults and 49 children) who lived near the Fukushima Daichi nuclear power station at the time of the accident.

Assuming that the whole-body absorbed dose of a resident in the initial year, *D(0)*, was well known through in situ measurements at the respective site, the chronological change of annual absorbed dose rate at time = t from radiocesium deposited onto the ground can be calculated as follows:

$$D(t) = (D(0) - D_{BKG}) \cdot \exp(-\lambda_{137} \cdot t) \cdot f(t) \tag{1}$$

where D_{BKG} is the background dose rate at the site; λ_{137} is the physical decay constant of ^{137}Cs; and *f(t)* is the attenuation function which is related to the cesium mobility in the surface soil layer. As the value of D_{BKG}, 0.05 µSv h^{-1} (0.44 mSv y^{-1}) was employed from a previous study on the external dose assessment of the evacuee in Fukushima Prefecture [19]. It is empirically known that the observed dose-rate changes can be approximated well with *f(t)* composed of a sum of two exponential functions, i.e., short-term and long-term decay components, as follows [2,20–23]:

$$f(t) = \alpha \cdot exp\left(-\frac{ln2}{T_S} \cdot t\right) + (1 - \alpha) \cdot exp\left(-\frac{ln2}{T_L} \cdot t\right) \tag{2}$$

where α is a fraction of the short-term component; T_S and T_L are environmental half-lives for short-term and long-term decay components, respectively, that characterize the disappearing rates from the respective local environment due to the natural removal phenomena and human activities for reusing the land. It is expected that the values of site-specific T_S and T_L could be derived from the on-site radiation monitoring data obtained for initial days by using the areal monitoring posts, monitoring vehicles, portable detectors, monitoring airplane, etc., as demonstrated at the Fukushima Daiichi accident [24].

In the present analyses, the environmental half-life of the long-term decay component (T_L) was assumed to be 50 years in reference to the relevant work of Golikov et al. [20] in regard to the Chernobyl accident and also the assessment of United Nations Scientific Committee on the effects of atomic radiation (UNSCEAR) regarding the Fukushima Dai-ichi accident [2]. As to the fraction (α) and environmental half-life (T_S) of the short-term component, large variations depending on the land-surface properties have been observed in the area affected by the Fukushima Daiichi accident [3,22,23]. Here, the 5th percentile, median and 95th percentile values of the "other than forest" undisturbed area in the evacuation zone were chosen from the paper of Kinase et al. [22] and subjectively formed three land categories, "Fast", "Middle", and "Slow", covering the wide range of the observed decaying patterns of ambient dose rate. The values adopted for α and T_S corresponding to these three categories are shown in Table 2.

Table 2. The values of environmental half-life (*Ts*) and fraction (α) of the short-term decay component; they were adopted from the 5th percentile, median and 95th percentile values derived from the observations of ambient dose in Fukushima Prefecture [22].

Land Category	T_S [y]	α [adu]
Fast	0.22	0.67
Middle	1.03	0.51
Slow	2.69	0.35

2.2. Projection of Cancer Risk

In cancer risk projections, the author focused on female infants (0-year-old children) because this group is known to be the most radiation sensitive for whole-body external exposure in regard to the all-cancer risk. Figure 1 shows the plots of the lifetime attributable risk (LAR) of cancer mortality attributing to a single whole-body exposure of 0.1 Gy as a function of age at exposure [10]; here, the LAR values have been converted to the fraction [%], while the original LAR values were given as the numbers of cases per 100,000 subjects. According to these data, the radiation-induced cancer risks of children is considered to be 2 to 3 times higher than those of adults; and female people are more susceptible to radiation than male people.

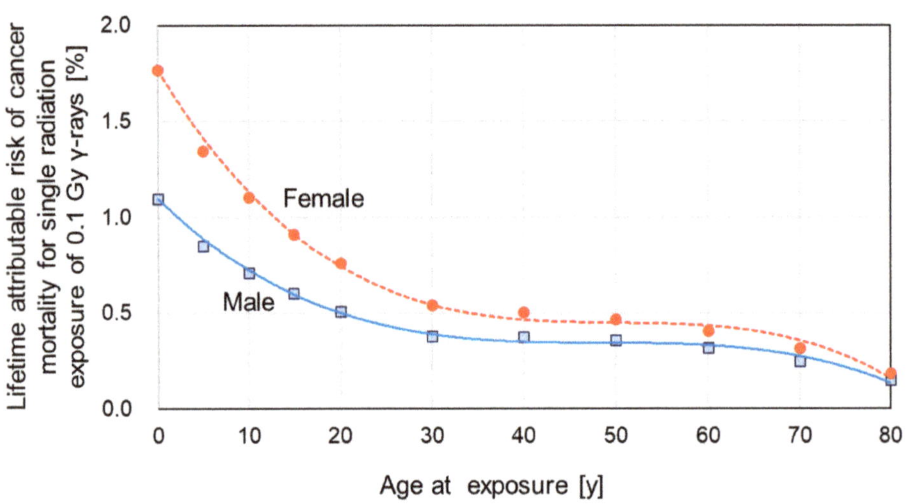

Figure 1. Lifetime attributable risk of radiation-induced cancer mortality as a function of age at exposure for male (solid line) and female people (dotted line) (after the BEIR VII report [10]).

With these facts, the author focused on female people as the critical group in assessment of life-long radiological risk and then predicted long-term cancer risks of female people for three time periods: 0 to 6 years (preschool), 0 to 18 years (up to high school graduation), and 0 to 80 years (lifetime). The integrated cancer-mortality risk for the period from 0 year to age y, $R(y)$, was calculated by integration of the products of the annual absorbed dose and corresponding LAR value for the respective period as follows:

$$R(y) = \sum_{t=1}^{y} \{1 - R(t-1)\} \cdot \frac{D(t)}{10^2} \cdot LAR(t) \tag{3}$$

where $R(0)$ is 0; and $D(t)$ is the annual absorbed dose rate at time = t [mGy y^{-1}]. In general, $D(0)$ is to be determined from on-site monitoring data; and $D(1)$ and subsequent annual dose rates are estimated by the method described in Section 2.1. Calculations of the $R(y)$ values were performed with a manually made program of Visual Basic for Applications (VBA) of Microsoft Excel.

3. Results and Discussions

3.1. Long-Term Cancer Risk

Figure 2 shows the chronological changes of annual external dose rates of γ-rays from ^{137}Cs deposited onto the ground classified by the land categories (Fast, Middle, and Slow). Clear differences were seen in the decay patterns of the dose rates among three categories, which suggests that the land-surface properties could bring significant effects on the long-term radiation exposures of residents.

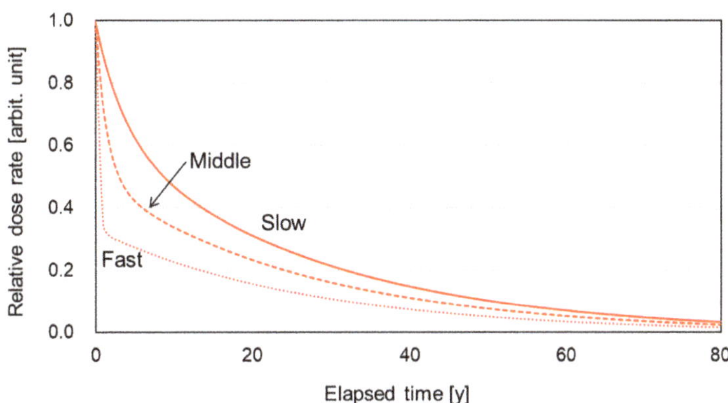

Figure 2. Predicted time change of the relative external dose rates from ^{137}Cs deposited onto the ground surfaces having different properties.

Using the predicted dose rate changes (Figure 2) and the data of age-dependent LAR (Figure 1), the author calculated integrated cancer-mortality risks of a female infant (0 years old) for her lifetime (up to 80 years old). The results are shown in Figure 3 as time plots of the integrated cancer risks for three land categories and also in Table 3 as the land-dependent integrated risks for three periods from the birth: 6 years (preschool period), 18 years (up to high school), and 80 years (lifetime). It was found that the cancer-mortality risk of a female person could change by a factor of 2 depending on the land-surface properties; in other words, where to live.

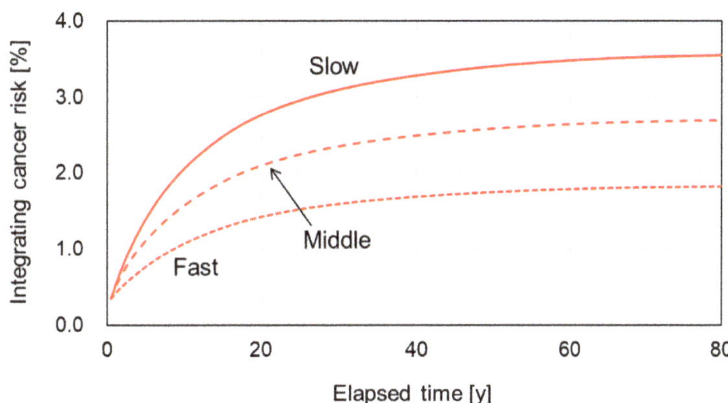

Figure 3. Predicted integrated cancer-mortality risk induced by γ-rays from ^{137}Cs deposited onto ground for a female infant (0 years old) at the initial dose rate of 20 mGy y^{-1}.

Table 3. Integrated cancer risks of a female person for three periods at the areas having different land-surface properties with the initial dose rate of 20 mGy y^{-1}.

Land Category	Integrated Cancer Risk [%]		
	~6 Years	~18 Years	~80 Years
Fast	0.88	1.40	1.83
Middle	1.28	2.05	2.69
Slow	1.64	2.70	3.56

3.2. Comparison with Nominal Risk

Effective dose is a widely used quantity to evaluate the stochastic health effects of radiation exposure for the aim of radiological protection. This quantity is not defined for an individual person but for "Reference Person" which represents a nominal population having typical age and sex distributions. On the basis of this concept, nominal risk coefficients for estimating nominal risks from the effective dose are given for two groups: whole population (the public) and adults (workers) [7]. Thus, the risks estimated from effective doses should be the same for all members of an intended group regardless of the different physical and physiological properties. This nature was intentionally given for avoiding unnecessarily discriminatory actions, which means that this quantity is unsuitable for the purpose of assessing individual risks [7,13].

If we employed the ICRP-recommended nominal risk coefficient (5.5% for cancer mortality) [7] in calculation of the cancer risk for the whole population of residents that were assumed to live in the land having an average land-surface properties (i.e., the Middle category), the nominal risks following the initial dose rate of 20 mGy y^{-1} were calculated to be 0.45% for the preschool age (~6 years), 0.86% for up to the high-school graduation, and 1.5% for lifetime (~80 years) (Table 4). These values were much lower (35% to 57%) than those calculated for a female person of the corresponding time period (Table 3).

Table 4. Nominal cancer risks estimated by using the nominal risk coefficients of ICRP [6] and the ratios of those nominal risks to the specific risks of a female person calculated in this study (Table 3) for three areas having different land-surface properties.

Land Category	Nominal Cancer Risk [%]			Ratio (Nominal/Specific)		
	~6 Years	~18 Years	~80 Years	~6 Years	~18 Years	~80 Years
Fast	0.305	0.582	1.03	0.347	0.417	0.562
Middle	0.449	0.861	1.53	0.353	0.421	0.566
Slow	0.584	1.15	2.03	0.357	0.426	0.571

The findings above imply that the currently common approach using effective dose would lead to an inappropriate judgement in the process of decision making regarding protective actions such as evacuation, relocation and decontamination following the concept for optimization in radiological protection, i.e., "As Low As Reasonably Achievable (ALARA)" [7,9]. It is expected that the alternate risk-based approach presented in this study would be useful for more appropriate judgement in optimization based on the integrated risk of all hazardous materials including heavy ions and toxic organic compounds in the living environment and the different sensitivities of the respective people to those materials.

It should be noted that difficulties still remain in the application of the presented approach to the decision making regarding protective actions. It is envisaged that many of the residents would prefer to behave as a community and they would not consider the difference in individual health status as an important matter. Moreover, the fear of radiation exposure would force them to just stay distant from the affected area, while it would not necessarily reduce the total health risk because some other effects, including psychological stress, could bring about more significant impacts on physiological status, as highlighted in the Fukushima Daiichi accident [5,6]. More efforts are needed for practical application of the presented approach based on the firm concept of minimizing the overall risk with consideration of social, economic, and psychological factors as well as the radiological consequences.

4. Conclusions

When a severe nuclear accident occurs, it is crucial to make prompt decisions on the measures for protecting people from excessive exposure to the radioactive sources in the living environment and other possible hazards. For contributing to this important decision making, the author has presented an approach for site-specific cancer risk projections of the most radiosensitive group following large-scale land contamination caused by radiocesium deposition. The obtained results indicated that accurate site-specific information on the behavioral characteristics of the radionuclides in the ground surface of the land is critically important for precise prediction of future consequences on the health of affected people.

The findings of this study indicate that we need to know, in advance, the physicochemical properties of the living environment of residents for maximizing our overall well-being through the best possible optimization of various risks and benefits for an unexpected event accompanied by large-scale environmental contamination. While recognizing that it is not an easy task, the author believes that such site-specific assessment of individual risks would enable the authority to make more appropriate decisions on protective actions, such as evacuation, relocation and decontamination, for averting excessive exposures to all hazards, including social, economic, and psychological impacts.

Funding: This research received no external funding.

Institutional Review Board Statement: Not applicable.

Informed Consent Statement: Not applicable.

Data Availability Statement: Not applicable.

Acknowledgments: This work was partially supported by the Program of the Network-Type Joint Usage/Research Center for Radiation Disaster Medical Science of Hiroshima University.

Conflicts of Interest: The author declares no conflict of interest.

References

1. Fukushima Prefectural Government. Fukushima Revitalization Station. Available online: http://www.pref.fukushima.lg.jp/site/portal/list271.html (accessed on 30 March 2021). (In Japanese)
2. United Nations Scientific Committee on the Effects of Atomic Radiation (UNSCEAR). *UNSCEAR 2013 Report to the General Assembly with Scientific Annexes, Annex A: Levels and Effects of Radiation Exposure due to the Nuclear Accident after the 2011 Great East-Japan Earthquake and Tsunami*; United Nations: New York, NY, USA, 2014.
3. International Atomic Energy Agency (IAEA). *The Fukushima Daiichi Accident*; IAEA: Vienna, Austria, 2015.
4. United Nations Scientific Committee on the Effects of Atomic Radiation (UNSCEAR). *UNSCEAR 2020 Report to the General Assembly with Scientific Annexes, Annex B (Advance Copy): Levels and Effects of Radiation Exposure due to the Accident at the Fukushima Daiichi Nuclear Power Station: Implications of Information Published since the UNSCEAR 2013 Report*; United Nations: New York, NY, USA, 2021.
5. Hasegawa, A.; Ohira, T.; Maeda, M.; Yasumura, S.; Tanigawa, K. Emergency Responses and Health Consequences after the Fukushima Accident; Evacuation and Relocation. *Clin. Oncol.* **2016**, *28*, 237–244. [CrossRef] [PubMed]
6. Maeda, M.; Oe, M. Mental health consequences and social issues after the Fukushima Disaster. *Asia Pacif. J. Public Health* **2017**, *19*, 365–465. [CrossRef] [PubMed]
7. International Commission on Radiological Protection (ICRP). *2007 Recommendations of the International Commission on Radiological Protection*; Publication 103; ICRP: Ottawa, ON, Canada, 2007.
8. International Commission on Radiological Protection (ICRP). *Radiological Protection of People and the Environment in the Event of a Large Nuclear Accident*; Publication 146; SAGE: London, UK, 2020.
9. International Commission on Radiological Protection (ICRP). *1990 Recommendations of the International Commission on Radiological Protection*; Publication 60; Pergammon Press: Oxford, UK, 1991.
10. U.S. National Research Council. *Health Risks from Exposure to Low Levels of Ionizing Radiation, BEIR VII Phase 2*; National Academies Press: Washington, DC, USA, 2006.
11. United Nations Scientific Committee on the Effects of Atomic Radiation (UNSCEAR). *UNSCEAR 2006 Report to the General Assembly with Scientific Annexes, Annex A: Epidemiological Studies of Radiation and Cancer*; United Nations: New York, NY, USA, 2008.
12. United Nations Scientific Committee on the Effects of Atomic Radiation (UNSCEAR). *UNSCEAR 2013 Report to the General Assembly with Scientific Annexes, Annex B: Effects of Radiation Exposure of Children*; United Nations: New York, NY, USA, 2013.
13. International Commission on Radiological Protection (ICRP). *Use of Dose Quantities in Radiological Protection*; Publication 147; SAGE: London, UK, 2021.
14. Howard, B.J.; Fesenko, S.; Balonov, M.; Pröhl, G.; Nakayama, S. A comparison of remediation after the Chernobyl and Fukushima Daiichi accidents. *Radiat. Prot. Dosim.* **2017**, *173*, 170–176.
15. Shizuma, M.; Nursal, W.K.; Sakura, Y. Long-term monitoring of radiocesium concentration in sediments and river water along five rivers in Minami-soma City during 2012–2016 following the Fukushima Daiichi nuclear power plant accident. *Appl. Sci.* **2018**, *8*, 1319. [CrossRef]

16. Wai, K.-M.; Krstic, D.; Nikezic, D.; Lin, T.-H.; Yu, P.K.N. External Cesium-137 doses to humans from soil influenced by the Fukushima and Chernobyl nuclear power plants accidents: A comparative study. *Sci. Rep.* **2020**, *10*, 7902. [CrossRef] [PubMed]

17. United Nations Scientific Committee on the Effects of Atomic Radiation (UNSCEAR). *UNSCEAR 2008 Report to the General Assembly with Scientific Annexes, Annex D: Health Effects due to Radiation from the Chernobyl Accident*; United Nations: New York, NY, USA, 2011.

18. Kim, E.; Kurihara, O.; Kunishima, N.; Nakano, T.; Tani, K.; Hachiya, M.; Momose, T.; Ishikawa, T.; Tokonami, S.; Hosoda, M.; et al. Early intake of radiocesium by residents living near the TEPCO Fukushima Dai-Ichi Nuclear Power Plant after the accident. Part 1: Internal doses based on whole-body measurements by NIRS. *Health Phys.* **2016**, *111*, 451–464. [CrossRef] [PubMed]

19. Akahane, K.; Yonai, S.; Fukuda, S.; Miyahara, N.; Yasuda, H.; Iwaoka, K.; Matsumoto, M.; Fukumura, A.; Akashi, M. NIRS external dose estimation system for Fukushima residents after the Fukushima Daiichi NPP accident. *Sci. Rep.* **2013**, *3*, 1670. [CrossRef] [PubMed]

20. Golikov, V.Y.; Balonov, M.I.; Jacob, P. External exposure of the population living in areas of Russia contaminated due to the Chernobyl accident. *Radiat. Environ. Biophys.* **2002**, *41*, 185–193. [CrossRef] [PubMed]

21. Tsubokura, M.; Murakami, M.; Nomura, S.; Morita, T.; Nishikawa, Y.; Leppold, C.; Kato, S.; Kami, M. Individual external doses below the lowest reference level of 1 mSv per year five years after the 2011 Fukushima nuclear accident among all children in Soma City, Fukushima: A retrospective observational study. *PLoS ONE* **2017**, *12*, e0172305. [CrossRef] [PubMed]

22. Kinase, S.; Takahashi, T.; Saito, K. Long-term predictions of ambient dose equivalent rates after the Fukushima Daiichi nuclear power plant accident. *J. Nucl. Sci. Technol.* **2017**, *54*, 1345–1354. [CrossRef]

23. International Atomic Energy Agency (IAEA). *Environmental Transfer of Radionuclides in Japan following the Accident at the Fukushima Daiichi Nuclear Power Plant*; IAEA-TECDOC-1927; IAEA: Vienna, Austria, 2020.

24. Nuclear Regulation Authority (NRA). Monitoring Information of Environmental Radioactivity Level. Available online: http://radioactivity.nsr.go.jp/ja/list/564/list-1.html (accessed on 10 May 2021). (In Japanese)

Review

The Processing of Calcium Rich Agricultural and Industrial Waste for Recovery of Calcium Carbonate and Calcium Oxide and Their Application for Environmental Cleanup: A Review

Virendra Kumar Yadav [1,*], Krishna Kumar Yadav [2,3,*], Marina M. S. Cabral-Pinto [4,*], Nisha Choudhary [5], Govindhan Gnanamoorthy [6], Vineet Tirth [7,8], Shiv Prasad [9], Afzal Husain Khan [10], Saiful Islam [11] and Nadeem A. Khan [12]

1 School of Life Sciences, Jaipur National University, Jaipur 302017, Rajasthan, India
2 Institute of Environment and Development Studies, Bundelkhand University, Kanpur Road, Jhansi 284128, UP, India
3 Faculty of Science and Technology, Madhyanchal Professional University Ratibad, Bhopal 462044, M.P, India
4 Geobiotec Research Centre, Department of Geoscience, University of Aveiro, 3810-193 Aveiro, Portugal
5 School of Nano Sciences, Central University of Gujarat, Gandhinagar 382030, Gujarat, India; nishanaseer03@gmail.com
6 Department of Chemistry, Guindy Campus, University of Madras, Chennai 600025, Tamil Nadu, India; gnanadrdo@gmail.com
7 Mechanical Engineering Department, College of Engineering, King Khalid University, Abha 61411, Saudi Arabia; vtirth@kku.edu.sa or v.tirth@gmail.com
8 Research Center for Advanced Materials Science (RCAMS), King Khalid University, Guraiger, Abha 61413, Asir, Saudi Arabia
9 Centre for Environment Science and Climate Resilient Agriculture, Indian Agricultural Research Institute, New Delhi 110012, India; shiv_drprasad@yahoo.co.in
10 Civil Engineering Department, College of Engineering, Jazan University, Jazan 114, Saudi Arabia; ahkhan@jazanu.edu.sa
11 Civil Engineering Department, College of Engineering, King Khalid University, Abha 61411, Asir, Saudi Arabia; sfakrul@kku.edu.sa
12 Department of Civil Engineering, Jamia Millia Islamia, New Delhi 110025, India; er.nadimcivil@gmail.com
* Correspondence: virendra.yadav@jnujaipur.ac.in (V.K.Y.); envirokrishna@gmail.com (K.K.Y.); marinacp@ua.pt (M.M.S.C.-P.)

Citation: Yadav, V.K.; Yadav, K.K.; Cabral-Pinto, M.M.S.; Choudhary, N.; Gnanamoorthy, G.; Tirth, V.; Prasad, S.; Khan, A.H.; Islam, S.; Khan, N.A. The Processing of Calcium Rich Agricultural and Industrial Waste for Recovery of Calcium Carbonate and Calcium Oxide and Their Application for Environmental Cleanup: A Review. *Appl. Sci.* **2021**, *11*, 4212. https://doi.org/10.3390/app11094212

Academic Editors: Raffaele Marotta and Bart Van der Bruggen

Received: 26 January 2021
Accepted: 9 March 2021
Published: 6 May 2021

Publisher's Note: MDPI stays neutral with regard to jurisdictional claims in published maps and institutional affiliations.

Abstract: Every year a million tonnes of calcium rich agro and industrial waste are generated around the whole globe. These calcium rich waste like finger citron, shells of cockle, mussel, oysters etc., and egg shell are biological sources which have various organic compounds. The inorganic calcium rich waste includes gypsum, dolomite, sludge etc., which are produced in surplus amount globally. Most of these by-products are mainly dumped, while few are used for land-filling purposes which leads to the pollution. These agro and industrial by-products could be processed for the recovery of calcium carbonate and calcium oxide particles by physical and chemical method. The recovery of calcium carbonate and calcium oxide particles from such by products make them biocompatible. Moreover, the products are economical due to their synthesis from waste materials. Here, in this current review work we have emphasized on the all the calcium rich agro industries and industrial by products, especially their processing by various approaches. Further, we have also focused on the properties and application of such calcium carbonate and oxide particles for the remediation of organic and inorganic pollutants from the environments. The recovery of such particles from these byproducts is considered not only economical and eco-friendly but it also minimizes the pollution present in the form of solid waste.

Keywords: calcium carbonate nanoparticles; calcium oxide nanoparticles; incense sticks ash; fly ash; sludge; eggshell

1. Introduction

Every day we encounter various agricultural and industrial products for our easiness and requirements. Out of these products, various products have different origin and chemical values. Agricultural industries is the backbone of any country, due to its importance in any countries GDP [1,2]. In addition, there are several other industries like steel, mineral processing, paper, and pulp industries etc., which also play a major role in a human's life. Both these types of industries fulfills our daily requirements. The progress of any country is measured by the growth of their industries. However, one major drawback with such industries is that they produces million tonnes of by-products every day throughout the whole world [3]. The accumulation of byproducts from such industries leads to the various forms of pollution due to the presence of toxic constituents [4]. The agricultural industry waste like citrus peel waste, paper and pulp waste contain moisture and attracts the flies and bugs leading to the health related issues [5]. While, the industrial waste like fly ash by-products of thermal power plants [6], gypsum waste from medical and hospitals leads [7] to the accumulation of hazardous by products due to the presence of heavy metals, etc. Here, we have focused on the recycling of calcium rich agricultural and industrial waste for the synthesis of value added minerals like calcium carbonate and calcium oxide. Both of these minerals are biocompatible, non-toxic, and used every day in oils, plastic, imaging purposes, alloys and catalyst [8], environment-friendly items [9], calcium-enriched food [10,11], drug delivery [12,13], cosmetics and pharmaceuticals [10], templates for microcapsules [14,15], and bone filling material [12]. Currently the commercial production of $CaCO_3$, calcium oxide and other derivatives of calcium is mainly carried out by using quick lime as a precursor material. Most of these quick limes are obtained by the mining which is laborious, and energy intensive process hence the final product is comparatively expensive [16]. So, there is an immediate need of a sustainable, economical, and an alternative source of calcium rich materials. There are several materials in our environment which are rich source of $CaCO_3$ like limestone, marbles, gypsum waste, cockle shell, eggshell, incense sticks, and high calcium fly ash (class C) and hard shell of shellfish, i.e., marine organisms (snails, oysters, shrimps, and pearls) [17–19].

Calcium carbonate ($CaCO_3$) is a white color powder and water insoluble-inorganic biomaterials [18]. In nature, $CaCO_3$ is present in three different forms, i.e., calcite, aragonite, and vaterite [20]. Due to its biocompatible nature, aragonite receives extensive research attention worldwide. Among all these three polymorphs, calcite is thermodynamically most stable [21]. Calcite is the most stable polymorph of $CaCO_3$ [22], but aragonite has higher density and hardness from the other two polymorphs of calcium carbonate, which makes them very suitable material in plastic, paper, glass, fiber, and other industry [23]. A study reported the needle-shaped aragonite synthesis under a controlled injection of CO_2 in $Ca(OH)_2$ slurry suspended $MgCl_2$ aqueous solution. These investigators also reported that the residual $MgCl_2$ solution from the previous step, could be reused for the synthesis of aragonite synthesis as both Mg^{2+} and Ca^{2+} ions are not incorporated into the aragonite crystals. At the time of formation of aragonite, Mg^{2+} ions acts as an impurity, which promotes the formation of aragonite crystals along with simultaneous impeding the nucleation and growth of calcite crystal. While another study carried by a group of investigators reported that the more the concentration of Ca^{2+} ions in the solution, the higher the possibility of formation of aragonite crystals will be [24–26]. To rationalize this cognition, numerous theories and empirical relationships have been provided such as alteration of surface charge, inhibition of calcite nucleation, and calcite poisoning model [21].

As an adsorbent calcium carbonate and calcium oxide particles have several advantages over the conventional adsorbents due to their low cost, biocompatibility, biodegradability, easy availability, non-toxic nature, and diverse polymorphs [27]. Due to all these properties calcium carbonate and calcium oxide particles have gained a huge attention in the field of environmental remediation especially for wastewater treatment. These calcium based minerals remediate the inorganic and organic pollutants from the environment by the adsorption process. The low cost and eco-friendly nature of calcium based minerals reduce the expenditure of the whole adsorption process.

2. Types and Properties of Calcium Carbonate Particles

In general, there are two sources of $CaCO_3$, one is ground calcium carbonate (GCC) and another one is precipitated calcium carbonate (PCC) [8], chalk, or marble. While the PCC is present as crystals and exists in three polymorphs, i.e., calcite, aragonite, and vaterite. Though they have several identical properties still they vary in the following manner as shown in Table 1. While all the PCC types of polymorphs are differentiated above in the Table 2.

Table 1. Differences between ground and precipitated calcium carbonate particles.

S. No.	Ground Calcium Carbonate	Precipitated Calcium Carbonate	References
1. Source	Extracted from earth	Present as crystals in calcite (rhombohedral), aragonite (orthorhombic), and vaterite (hexagonal) forms	[28]
2. Examples	Examples: chalks, marble	-	
3. Processing	Grinding is required either in wet or dry conditions.	-	
4. Available in market	-	Commercial PCC was produced in 1841.	
5. Methods for synthesis	Thermolysis	Three common processes for the production of synthetic PCC(1) lime soda process(2) calcium chloride process(3) the carbonation process	[29]

Table 2. Differences between different types of calcium carbonate nanoparticles polymorphs.

S. No.	Calcite	Aragonite	Vaterite	References
1. ThermalStability	Thermodynamically most stable	Moderately stable	Least stable polymorph	[21]
2. Solubility	-	More soluble and denser than calcite	-	
3. Structure	Exists as a trigonal crystalline form in nature	Forms needle-like orthorhombic crystals and formed at higher temperature and pressure	Hexagonal structure and rarely seen in the natural mineral	
4. Stability	-	Metastable and slowly gets converted to calcite		[23]
5. Interaction with water and reorganization	-	-	Vaterite when exposed to H_2O, slowly dissolves and recrystallizes to a stable form	
6. Porosity, surface area	-	-	Large porosity and large surface area	
7. Disintegrity	-	-	Rapid disintegration under relatively mild conditions	
8. Biocompatibility	-	Biocompatible properties		[21]

Calcium based particles has drawn their attention towards different scientific filed in the last decade due to their diversity in morphology, biocompatible in nature, economical, and non-toxic effect on the environment [27]. Since the calcium is already present in our body, microbes, plants, etc., it does not have a detrimental effect on any of these species, so it is considered as superior particles, in comparison to other metal oxide particles. Such as, ZnO, TiO_2, etc., are photo catalytic in nature, whereas CuO, CoNPs, etc., are toxic due to their heavy metal nature. These metals oxide particles persists in nature for much longer time and may show, bio magnification in the higher animals in the food chain. Whereas calcium is already present in almost most of the organisms, so there is already degradation mechanism present in the organism for calcium based particles. The types of calcium carbonate particles are based on their morphology. Calcium carbonate particles (CCPs) are white color powder, insoluble in water but soluble in HCl, biocompatible, pH sensitive, biodegradable, abundant in nature, economically viable, and exhibit polymorphism [27]. CCPs aggregate easily during the process of preparation and disposal [30].

3. Different Sources of Calcium Carbonate Particles (CCPs)

In nature, there are numerous calcium-rich particles which can be used as a source of CCPs synthesis. For instance, there are biological materials like cockle and shellfish. All these sea animals have a hard covering on their surface, which is meant for protection, safety from the wear and tear, and prey. These animals have very soft internal organs, so a hard-calcareous covering is required. So, their shells can be used for the extraction of CCPs by various means of processing. Most of these sea animals lose their hard covering after their death and there is the mineralization of calcium into the seawater. In addition, there are several domestic and industrial waste which have also higher content of calcium. For instance, eggshell and incense sticks ash from the domestic waste. While high Ca fly ash, gypsum waste [30], calcium sulfide waste [31], sludge is industrial waste which is also rich sources of calcium. All these materials are discussed below in detail by emphasizing the steps for the recovery of CaO NPs. Most of these calcium-rich materials are waste so the recovery of CCPs and calcium derivatives from these materials are considered economical and environment-friendly.

3.1. Domestic Waste

The waste which is generated in the houses especially in the kitchen is called household waste, domestic waste, or kitchen waste [32]. Among all the high calcium-containing waste, incense sticks ashes are considered as most orphan and least considered; though it did produce about 1–2 MTs in India, and this figure would rise drastically if the major incense sticks consuming countries like China, USA, and South Asian countries were also considered. South Asian countries alone generates tonnes of ISA at religious places [33]. Though it is also produced in a small amount at various houses of South Asian counties [34]. Several investigators have classified the eggshell in household waste. However, the utilization of eggs at home is lesser than the industries so few investigators have categorized them into industrial waste. Therefore, here we will consider only incense sticks ash waste as a source of CCNPs.

3.2. Incense Sticks Ash

Incense sticks ash (ISA) is one of the major household wastes in South-east Asian countries like China, Taiwan, Thailand, and India. Incense sticks are majorly used in the temples, churches, mosques, houses, and other religious places. In Taiwan alone, a total of 3580 tons of incense sticks are consumed yearly in temples alone and if household burning is also considered then this value may reach to double or triple and may indicate an environment hazardous situation [33,35]. The size and composition of an incense sticks varies from one country and one religion to another. Currently, most of the countries, including India, do not have exact data on the incense sticks ash production. The only information we have regarding this is that: Brazil, China, U.S.A and India, are the leading manufacturer of incense sticks. India is the third largest manufacturers of incense sticks and fourth largest consumer of incense sticks. Along with USA, India also exports most of the incense sticks to Gulf countries, and Mexico. In India, the average size of an incense sticks is 8–9 inches and width is about 1–3 mm. Generally, Indian markets have two types of incense sticks: one is charcoal, or coal based (black in color), whereas the other one is non charcoal based whose color could be grey, or any other color. The black color incense sticks ashes have ferrous, silica, alumina, carbon, and calcium oxides. Indian incense sticks manufactures uses calcium phthalate in order to minimize the smoke released during burning of sticks. So, this is the one of the source of high amount of calcium oxides in the ISA. The ISA have nearly 50–56% calcium oxides [36]. The composition of incense are herbal and wood powder 21%, fragrant materials 35%, adhesive powder 11%, bamboo stick 33% by weight [37,38]. The complete combustion of a 9 inch and 1–2 mm incense stick, produces one-third ash by weight while the remaining 60–70% weight of incense sticks is made up of volatile matters like essential oil, fragrance, and combustible matter. Hazarika et al., 2018 and Yadav et al., 2020, reported that India alone consumes approximately 3–4 million tonnes of incense sticks, while USA was one of the largest importer of incense sticks from India. Based on the above information, we can predict that India alone generates 0.3–0.4 million tonnes of ISA from the various religious places and homes. The same information from other countries would be impossible to estimate as there is hardly any information available from the government or in the literature [36].

The chemical composition of incense sticks ash is given below in Table 3. The X-ray fluorescence analysis reveals that the incense ash has CaO 50–55%, Ferrous oxide (Fe_2O_3/Fe_3O_4) 4–5%, Aluminum oxide (Al_2O_3) 4–5%, Silica oxide 15–20%, magnesium oxide 4–5%. In addition, it also has a high amount of alkali oxides, i.e., Na_2O, K_2O, traces of TiO_2, CuO, and other heavy metals. Figure 1 shows the flowchart of the synthesis of $CaCO_3$ from ISA, while Figure 2 shows field emission scanning electron micrograph (FESEM) of calcium carbonate microparticles from incense sticks ash. Oral and Ercan (2018), also reported the calcium carbonate particles of various sizes synthesized by chemical routes. Out of these cuboidal shaped particles were also observed, by varying the temperature of synthesis [39]. In Figure 1, initially ISA was mixed with distilled water (1:5) to form a slurry. Further with the help of an external magnet, the ferrous particles were separated as ferrous will add impurity in the final calcium carbonate particles. Further, it was dried in an oven at 50 °C, followed by treatment with sulphuric acid at 90 °C for 90 min to extract alumina along with stirring at 400 rpm. Further, after completion of the reaction, mixture was cooled to room temperature (RT), followed by the centrifugation. The obtained residue was washed several times with distilled water, followed by drying in an oven at 60 °C. About 8 g of residue was treated with 4 M NaOH at 90 °C for 90 min along with stirring at 400 rpm. Further, the residue was calcinated at 600–700 °C for 6 h, followed by dilute HCl treatment in a round bottom (RB) flask, along with stirring at RT. Around 10 mL aqueous leachate obtained with HCl treatment was mixed with dropwise methanol in a RB flask at RT. The mixture was centrifuged at 7000 rpm for 10 min, where the residue was discarded while the supernatant was collected. Further, about 10 mL of supernatant was taken up, to which 2% solution of sodium bicarbonate was added drop wise along with stirring at RT in a RB flask. The mixture was centrifuged at 7000 rpm for 10 min, supernatant discarded while the precipitate was collected. Finally, it was washed twice each with distilled water and ethanol followed by drying at 60 °C for overnight.

Table 3. The elemental composition of incense sticks ash by X-ray fluorescence.

S. No.	Chemical Elements	Weight (%)
1.	CaO	50–55
2.	MgO	4–5
3.	SiO_2	15–20
4.	Al_2O_3	4–5
5.	Fe_2O_3/Fe_3O_4	4–5
6.	TiO_2	2–3
7.	Others ($CuO + Na_2O + K_2O$)	5–10

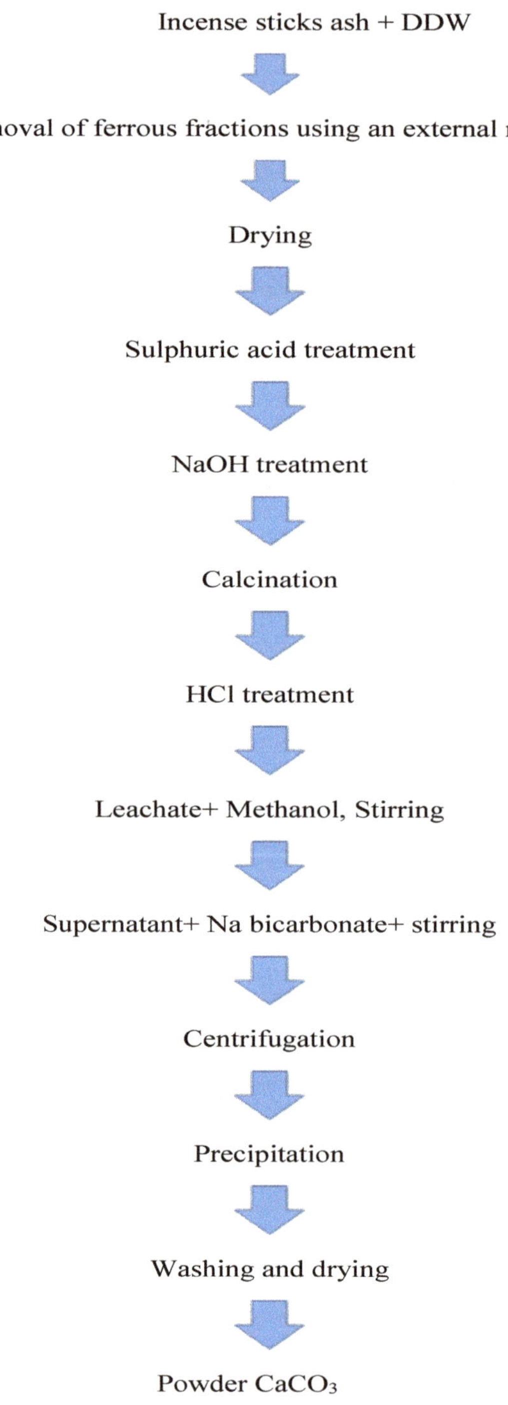

Figure 1. Flow chart for the separation of calcium carbonate from incense stick ash.

Figure 2. Field emission scanning electron micrograph (FESEM) of the calcite phase of calcium carbonate microparticles from incense sticks ash.

4. Extraction of Calcium Oxide Particles from Eggshell Waste

The bioconversion of waste material like eggshell from poultry industries into a valuable material is highly significant for economic development and waste management [40]. The eggshell waste is classified in both industrial wastes as well as in agricultural waste. The utilization of these wastes for the production of valuable materials like fertilizers, feed supplements, adsorbents, and CCNPs not only provides a solution to the disposal of eggshell but also minimizes the pollution in the environment [41,42]. Among all the countries, China, USA, and India is the leading producer of eggs which contributes near about 80 million metric tonnes, where Chinas share is 458, USA's share is 109, and India's share is 95 billion eggs per shell as per the 2017 literature. While in 2018, the global egg production was approximately, 78 million metric tonnes. These 78 million metric tonnes produced about 8.58 million metric tonnes of eggshells which are mainly dumped of as waste, while some parts of the world, it is used as landfills, and for fillers, etc. [43]. The eggshell consists of about 10–11% weight of the total weight of an egg. Annually around, 250,000 tons of eggshell waste is generated during processing [44]. There are 150,000 tons of chicken eggshell disposed of in landfills every year in the U.S alone. The eggshell comprises about 95% $CaCO_3$ as calcite and 5% organic materials such as type X collagen, sulfated polysaccharides, and other proteins which is provided below in Table 4 [45]. The disposal methods for waste eggshells are 26.6% as fertilizer, 21.1% as animal feed ingredients, 26.3% discarded in municipal dumps, and 15.8% used in other ways [41]. Habte et al., 2020 reported the synthesis of microsized (10–30 μm) aragonite calcium carbonate particles from eggshell waste by carbonation method along with calcination at high temperature.

Table 4. The chemical composition of chicken eggshell.

S. No.	Chemical Elements	Concentration (mg/L)
1.	Ca	2296–2304
2.	Mg	849–852
3.	Na	33–35
4.	K	16–19
5.	Fe	1.01–1.43
6.	Zn	0.95–1.03
7.	Cu	0.062–0.064

India is also one of the major poultry industry countries which have a poultry population of 489 million, producing 47 billion eggs per year. India ranks third among the highest egg producing countries in the world [46]. The outermost cover of the egg is called eggshell whose weight is 10–11% to that of the total weight of the egg and made up of mainly $CaCO_3$ (96%) and trace elements. The chemical composition of eggshell waste is given below in the table. Currently, it is mainly dumped as landfills in most of the developed or developing countries. Disposal of eggshell and the underlying membrane further contributes to environmental pollution [47].

Hassan et al. (2013) reported the synthesis of CCPs from the chicken eggshell waste, which encompasses the following steps cleaning and size reduction of eggshell, followed by surface modification by the sonochemical method for enhanced dispersion [48,49]. Hariharan et al. (2014) reported the successful synthesis of calcite NPs from eggshell waste using gelatin [50,51]. The calcite polymorph of $CaCO_3$ was synthesized using from chicken eggshells employing gelatin via precipitation method and the confirmation of the nano calcite was done by X-ray powder diffraction (XRD), Fourier-transform infrared spectroscopy (FTIR), UV–Visible (UV–Vis) spectroscopy and scanning electron microscopy (SEM). The identified particles were calcite polymorphs with a particle size of 25 nm. The analysis results confirmed the formation of calcite NPs and the obtained results were compared with $CaCO_3$ synthesis without using gelatin. Render et al. (2016) reported the synthesis of CCPs by using eggshell waste that has been used for enteric drug delivery. The sequential steps involved in the synthesis of CCPs from eggshell was initial cleaning, mechanochemical milling, and Sonochemical irradiation [52]. Pandita and Fulekar in 2017 and 2019 reported the synthesis of CCPs from eggshell waste [53]. The investigators, have reported the synthesis of micron and nano sized $CaCO_3$ and further transformed them into the CaO NPs by calcination at 900 °C for 4 h. They used these CaO NPs for increasing the efficiency of biodiesel by as micronutrient for the algae the microalgae. Further, Ahmad et al. (2020) has synthesized nano calcium oxide from the eggshell waste and utilized them for the direct transesterification of *Chlorella pyrenoidosa*. The eggshell was washed properly, crushed by mortar pestle, dried at 105 °C, and calcined at 870 °C for three hours. The final powder was analyzed by the sophisticated instruments, for their morphological properties. The size of CCPs was varying from 25 to 100 nm. Further they have used these nano CaO successfully up to six cycles for the transesterification [54]. Figure 3 demonstrates the flowchart of synthesis of $CaCO_3$ from egg shells.

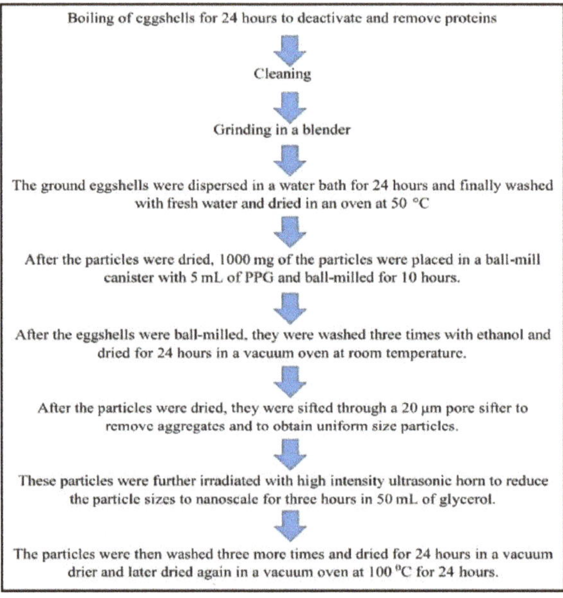

Figure 3. Steps involved in the synthesis of calcium carbonate nanoparticles from eggshell waste.

5. Biogenic Waste of Marine Organisms

5.1. Shells of Shellfish

Shellfish are sea animals that belong to the class invertebrates, having mainly two types of animals: crustaceans and mollusks. They mainly include lobsters, shrimps, crayfish, crab, and krill. They have a very hard covering on their external surface which consists of chitin. The shells of such marine organisms have a high content of $CaCO_3$ which is meant for protection from the external environment and predator. After the death of such marine organisms, there is biomineralization of $CaCO_3$ and ultimately calcium ion is added into the seawater [55]. Every year a million tonnes of sea foods are generated in the whole world and major contributors are countries with larger coastal area in the form of their boundary which favors the aquaculture. As far as India is concerned, till 2018, it occupies second rank in aquaculture and third rank in fisheries contributing 1.07% to the nations GDP. Major, waste of sea foods are heads, trimming residue, tails, shells, and scales. Out of these shell fish soft flesh part is consumed as food while the hard calcareous covering is considered as waste which is mainly dumped of as landfills. Numerous investigators have also reported the synthesis of CCPs from the shells of shellfish. Jaannah et al., 2018, reported the synthesis, characterization, and application of shellfish-derived CCPs for the antibacterial assay [17]. Initially, they synthesized the CCPs and formed a nanocomposite with MgO which was further assessed for their antimicrobial activity against *Escherichia coli* and *Staphylococcus aureus* bacteria. They have also reported the synthesis of $CaCO_3/MgO$ nanocomposite material by the optimum clear zone and able to utilize shell waste as an antibacterial ingredient of natural materials [17]. The schematic step involved in the synthesis of CCNPs from shellfish is given below in Figure 4.

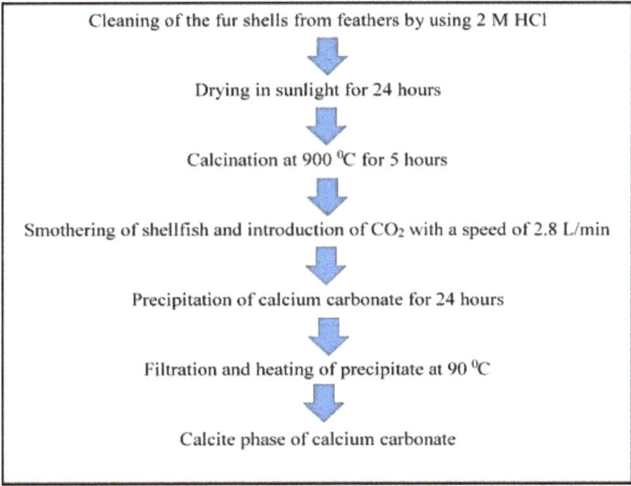

Figure 4. Schematic diagram of extraction of calcite phase from the shellfish.

5.2. Cockle Shells

As per the World Fisheries and Aquaculture, USA, 2020, report, about 17,510.9 thousand tonnes of mollusks produced in the 2018, out of which cockle were 433.4 thousand million tonnes, i.e., 2.5% share of mollusks. Which need to draw the attention towards the cockle shell management otherwise it will lead to the loss of economy in the form of disposal of these cockle shells. So, the recycling of such cockle shells, will not only provide an alternate source of calcium carbonate and their derivatives, but it will also minimize the solid waste. Numerous investigators have reported the recycling of cockle shells for value added materials like calcium oxides and calcium carbonates out of which most recent ones are cited below.

Majusha Hariharan et al. (2014) synthesized and characterized the nano calcite through the precipitation method from the cockle shells using chitosan as precursor materials. The schematic steps involved in the synthesis of CCNPs from cockle shell are given below in Figure 5. The nano calcite was characterized by SEM, XRD, UV–Vis, and FTIR for confirmation of the particles which was later on compared with the commercial nano calcite. It was found that cockle shells are a potential source of nano calcite, which was by-products of the seafood industry so the method was cheaper and environment-friendly [50].

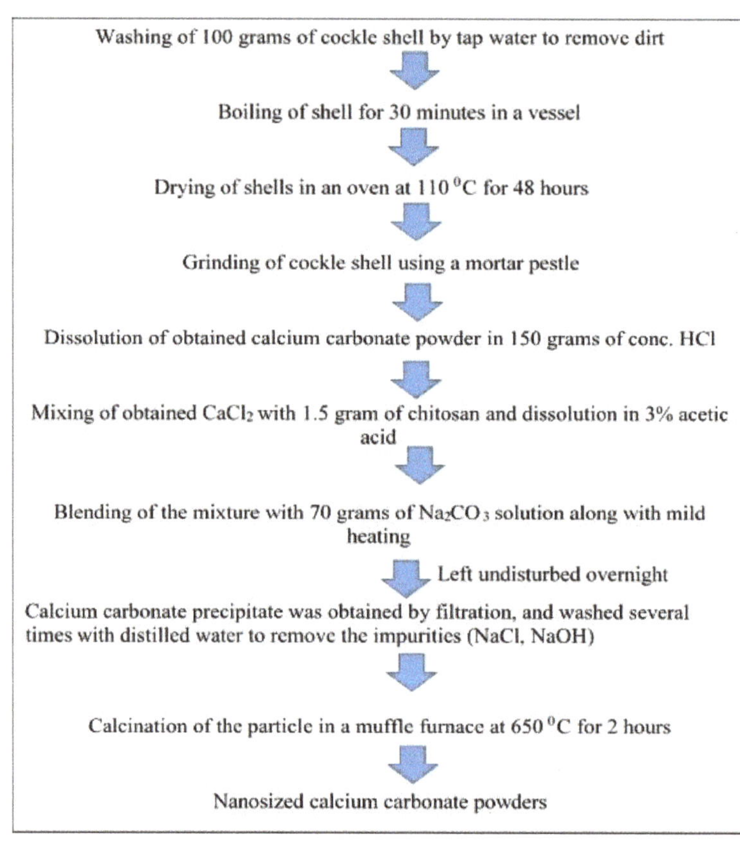

Figure 5. Steps involved in the synthesis of calcium carbonate nanoparticles from cockle shells.

Islam et al. (2012) reported an easy, cost-effective, and novel method for the synthesis of CCPs (aragonite) from the cockle shells whose schematic synthesis is given below in Figure 6. Aragonite is one of the least abundant biogenic polymorphs of $CaCO_3$ which is commonly used as a biomaterial for the repair of the fractured bone, development of advanced drug delivery systems, and tissue scaffolds. The size of obtained aragonite NPs size was 20 ± 5 nm with high purity which was confirmed by the variable pressure SEM, transmission electron microscopy (TEM), FTIR, thermogravimetric analyzer (TGA), XRD, and energy-dispersive X-ray spectroscopy [23].

5.3. From Oyster and Mussel Shells

Oysters and mussels are also one of the major sea foods consumed in most of the countries around the globe. Out of 17,510.9 thousand tonnes of mollusks produced in 2018, the mussels share was 1205.1 thousand tonnes, i.e., 6.9% which after processing generates tonnes of shells every day. So, the waste generated from such seafoods maybe requires attention towards recycling, otherwise their disposal will lead to the loss of economy [50]. Numerous investigators reported the synthesis of calcium carbonate and oxide from the oyster and mussel shells by various chemical methods out of which most recent ones are cited below.

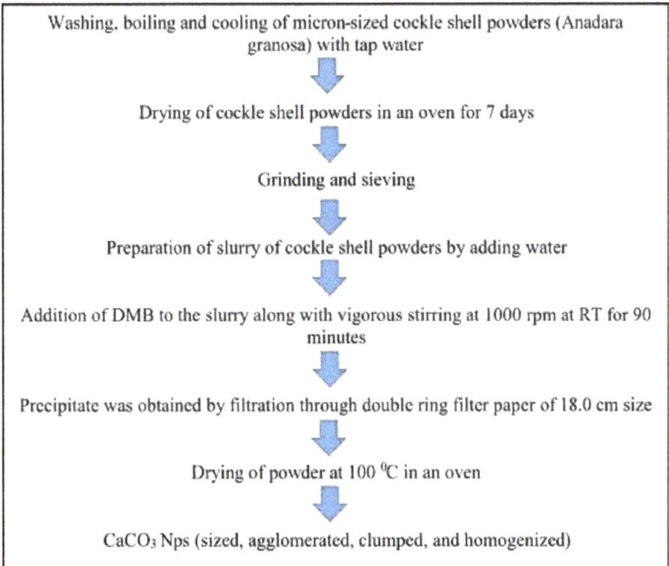

Figure 6. The schematic diagram for the synthesis of calcium carbonate nanoparticles from cockle shells.

Hamester et al. (2011) obtained calcium carbonate from mussel and oyster shells were further used as filler in polypropylene and compared their properties with polypropylene and commercial calcium carbonate composites [56]. For the synthesis of calcium carbonate, the mussel and oyster shells were heated in an oven at 200 °C for 1 h to make the shells more brittle and submitted to milling in a high-speed planetary mill with porcelain jar and alumina balls for 15 min with water. The obtained powders are again heated to 500 °C and maintained for 2 h. Further, to undo the clusters a new milling was performed without addition of water for 1 min. The resulted powders were characterized by a laser diffraction analyzer and X-ray fluorescence for chemical composition [57].

6. Recovery from Industrial Waste

Every year a million tons of industrial waste in the form of gypsum, calcium sulfide, sewage sludge, high calcium fly ash, wollastonite, etc., are produced around the whole world. Most of these are the byproducts of the industries which are generally dumped in the near vicinity of the industries or as landfill [58–60]. The dumping of these waste materials may invite mosquitos and other insects and may lead to various diseases. Moreover, most of these materials are processed materials that have a higher amount of toxic metals that may leach into the land or water bodies and may contaminate them. The aquatic animals may accumulate these toxic metals, i.e., bioaccumulation and lead to bio magnification. So, there is an urgent need to utilize these waste materials for the recovery of value-added materials like calcium. The recovery of calcium-based products from such waste will reduce pollution and provide an alternate source for the CaCO$_3$ particles via a cost-effective and eco-friendly method [61]. Some of the materials that have been used earlier for the synthesis of CCNPs are described below in detail.

6.1. Recovery of Calcium Oxide/Carbonate from Sludge

Sludge is one of the major wastes from houses and industries. Though it has numerous mixtures in it has a high amount of calcium. In a highly populous country like India, about 61,754 million liters per day sewage is generated, and 38,791 MLD is untreated sludge. Every year India produces 277 MTs of solid municipal waste according to estimate of 2016. Whereas this figure is projected to reach to 378.8 MTs in 2030 and 543.3 MTs by 2050. This figure is really need to focus the government and other bodies for recycling and recovery of value added materials from such waste [62,63]. The sludges are reported to have higher percentage of calcium makes it a suitable material for the extraction of CCNPs. Moreover, the raw material is again a waste that needs special treatment for disposal. So, the recovery of $CaCO_3$ from sludge not only reduces solid waste or pollution but also helps in producing valuable minerals from it [64]. The synthesis of $CaCO_3$ from sludge is a low-cost technique, eco-friendly. Till yet an only a countable number of works is done in this filed.

The recovery of CCNPs from the waste sludge by the flotation technique [65]. They investigated the effective dosage of floating agents such as sodium oleate and sunlight dish liquid) and the percent solids of the slurry. They have used floating sieved and un-sieved materials and $CaCO_3$ was estimated for both conditions as well as from tailings. Initial $CaCO_3$ analysis for the bulk material indicated that sieved and un-sieved materials had 63.4% and 32.9% $CaCO_3$ content by weight, respectively. Therefore, it was concluded that for un-sieved material sunlight dishwashing liquid was a better collector compared to the latter. The results proved that there is a great potential for recovering commercial-grade limestone from wastewater sludge.

6.2. Recovery of Calcium Carbonate/Oxide from Dolomite

Dolomites $[CaMg(CO_3)_2]$ are anhydrous carbonate minerals of calcium and magnesium carbonate [66]. It is widely used in the steel and industries. Such iron or steel based industries produces 40–70% byproducts including dolomite as waste. In general, dolomite is used for the preparation of precipitated CaCO3 by separating Ca and Mg fractions. The major problem in their separation from dolomite is that both the elements have a lower solubility in the water [67]. Some investigators have successfully used sucrose for the extraction of Ca and Mg from dolomite. When the calcinated dolomite is dissolved in a sucrose solution, then there is the conversion of CaO into calcium sucrate leaving unreacted Mg in the mixture as a precipitate [68]. When calcined dolomite (CaO.MgO) is dissolved in a sucrose solution, CaO will be converted into soluble calcium sucrate, while MgO remains unreacted and presents in the precipitated form [69]. Figure 7 illustrations the stepwise synthesis of CCPs from dolomite.

6.3. Synthesis of Calcium Oxide/Carbonate Particles from Gypsum Waste

Gypsum waste is a waste product of the reverse osmosis (RO) desalination process [70]. Gypsum is widely used for dental applications [71] which on disposal into the environment may challenge as a hazardous material. When such gypsum are landfilled into the environment, there is a reaction with biodegradable waste which may produce poisonous and odorous hydrogen sulfide gas [72,73]. Gypsum alone is widely used in the dentistry in hospitals and medical colleges. Every year one medical college produces 100–500 kg of gypsum waste which could vary based on the patient footfall per year. As per the literature, there are about 345 dental colleges in India, so approximately these colleges generates about 173 tonnes of gypsum waste. In addition, gypsum waste is also generated from various construction site, industries, and mineral processing industries. Gypsum waste can be thermally reduced into CaS, which is then subjected to a direct aqueous carbonation step for the generation of H_2S and $CaCO_3$. CaS can be successfully converted into $CaCO_3$; however, the reaction may yield low-grade carbonate products (99% as $CaCO_3$) or precipitated $CaCO_3$ can be developed and optimized [70].

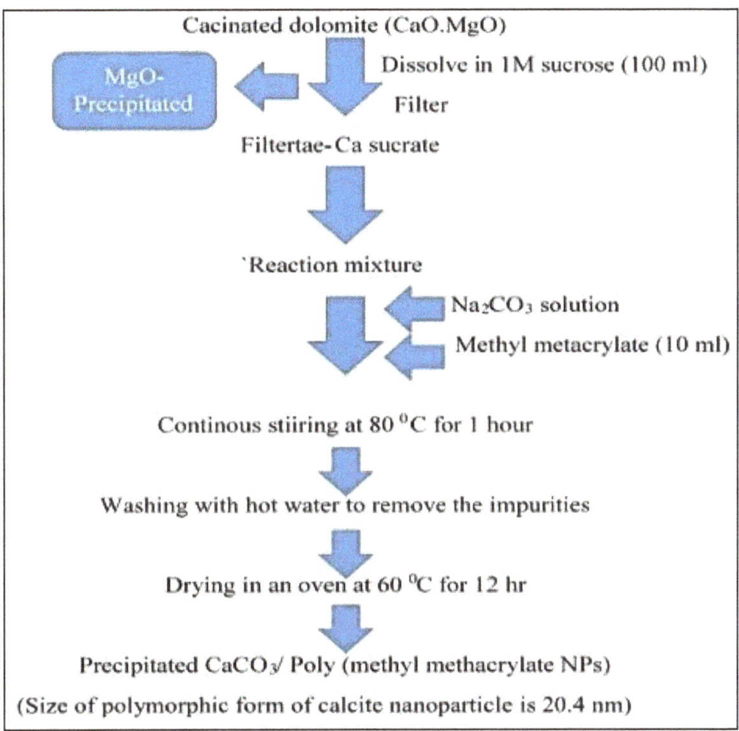

Figure 7. Steps involved in the synthesis of calcium carbonate nanoparticles from dolomite.

Beer et al. (2014) reported the synthesis of CCPs from the gypsum waste, where elemental sulfur was the by-product. Here the first step was the thermal reduction of gypsum waste into calcium sulfide (CaS) followed by its direct aqueous carbonation yielded low-grade carbonate products or precipitated $CaCO_3$. The process used an acid gas (H_2S) to improve the aqueous dissolution of CaS, which is otherwise poorly soluble. The carbonate product was primarily calcite (99.5%) with traces of quartz (0.5%). Calcite was the only $CaCO_3$ polymorph obtained; no vaterite or aragonite was perceived. The schematic steps involved in the synthesis of CCNPs from gypsum waste are shown below in Figure 8. The product was made up of micron-size particles, which were further characterized by XRD, TGA, SEM, Bruner–Emmett–Teller (BET), and true density. The batch recovery of $CaCO_3$ from waste gypsum slurry using sodium carbonate under ambient conditions that were further utilized for pre-treatment of acid mine drainage from coal mines [70]. US patent no 2013/0288887 A1 reported a simple, cost-effective, and novel method for the recovery of nano $CaCO_3$ from the gypsum waste slurry [74]. There is a lower decomposition temperature of $CaCO_3$. The synthesized nano $CaCO_3$ was used as a nano adsorbent for the adsorption of carbon dioxide, and as a complex catalyst for reactive sorption enhanced the reforming process for hydrogen production from methane. The CaO-based carbon dioxide adsorbent shows good cycle stability and fast sorption rate, and complex catalyst used for reactive sorption enhanced methane steam reforming can obtain the hydrogen with a purity of more than 90%.

The CCPs can be synthesized by a simple thermal decomposition method [75], where the dolomite is calcined at 800 °C, and it is decomposed into $CaCO_3$ and CaO in the presence of air. The major disadvantage with this method is that there are adsorption and precipitation of $AsCO_3$ and AsO [76]. Chilakala et al. (2016) reported an innovative convenient, and cost-effective carbonation method for the synthesis of pure aragonite needle phase of CCPs from dolomite [77]. The characterization of PCC was done by XRD and SEM for the morphological and mineralogical and aspect ratio (ratio of length to the diameter of the particles). The synthesis of PCC was carried out in two steps, at first, after calcinated, dolomite fine powder was dissolved in water for hydration, the hydrated solution was then mixed with an aqueous solution of magnesium chloride at 80 °C, and then CO_2 was bubbled into the suspension for three hours to produce aragonite PCC. Finally, aragonite type precipitated $CaCO_3$ can be synthesized from natural dolomite via a simple carbonation process, yielding a product with an average particle size of 30–40 μm [78].

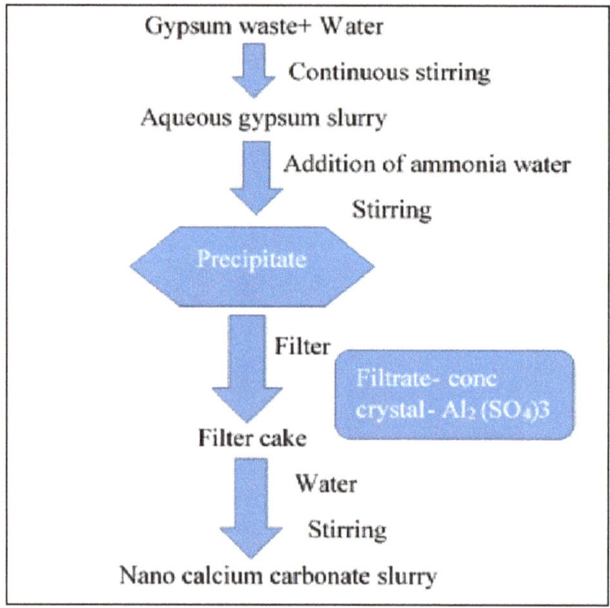

Figure 8. Steps involved in the synthesis of calcium carbonate nanoparticles from gypsum waste.

6.4. Recovery of Calcium Oxide/Carbonate Particles from Finger Citron Residue

About two-thirds of the world's citrus fruits are produced by Brazil, China, India, Mexico, Spain, and the USA. In 2016, about 124 Million gallons of citrus fruits were produced out of which 50–60% used as fresh fruits while the remaining 40–50% was subjected to industrial processing for extraction of juices, etc. Based on the applied technology and type of cultivars, the citrus peel waste of such industries varies from 50 to 70% w/w of processed fruits [79,80]. This leads to the generation of 10 Mg of citrus peel waste, which has moisture whereby it attracts flies, insects, and mosquitoes and leads to the pollution. So, it is important to process the citrus peel waste for the recovery of value added materials. One such citrus fruit is finger citrons (FG) whose shape is unusual and the fruit is segmented into finger-like sections [81]. Finger citron, which is a subtropical plant, is grown in China's Sichuan, Fujian, Guangdong, and Zhejiang provinces [82]. In the time of finger citron beverage processing, a huge amount of Finger citron residues (FCRs) are produced which is disposed of as garbage Such disposals sometimes may give rise to other environmental problems [83]. It has a porous texture and contains numerous organic compounds which makes them a preferable precursor material for the synthesis of porous materials that can also be potentially applied for environmental protection. They are almost similar to lemon and used as a substitute for lemon in China. FCR is the residual of the fruit after extraction of juices from it. The FCR is reported to have numerous valuable minerals and elements but it has a high amount of Ca which makes them a suitable material for the extraction of CCPs. The synthesis mechanism of CC NPs from FCR is given below in Figure 9.

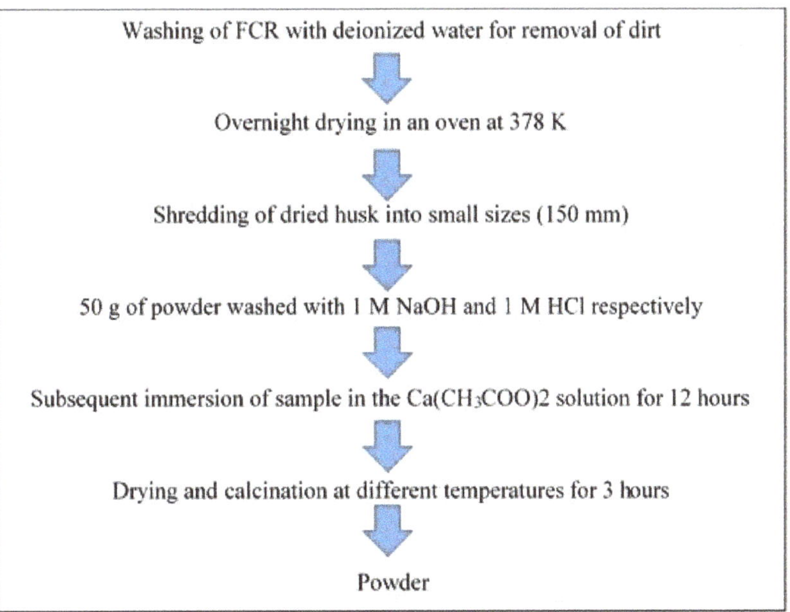

Figure 9. Steps involved in the synthesis of calcium carbonate nanoparticles from finger citron residue.

6.5. Recovery of Calcium Oxide/Carbonate from Waste Calcium Sulfide

Calcium sulfide (CaS) is one of the major waste of several industries which are based on coal [84]. This is a waste material that is produced in large amounts for example, gasification of 1000 tons/day of 3% sulfur coal produces 67 tons/day of CaS) during the desulfurization of hot coal gas by a limestone-based adsorbent. The main problem with such waste is that it is rich in sulfur so cannot be landfilled as there could be potential H_2S gas evolution and leaching of sulfur in the soil. Brooks and Lynn (1997) reported the synthesis of $CaCO_3$ and H_2S from the waste CaS where, the calcium sulfide is dissolved by reacting with H_2S forming a complexed either with aqueous methyl diethanolamine (MDEA) or other alkanolamines [85]. Further, in the very next stage, $Ca(HS)_2$, which is extremely soluble, reacts with CO_2 complexed with MDEA (aqueous). These precipitates the pure and uniformly sized crystalline $CaCO_3$ and to forms MDEA-complexed hydrogen sulfide.

6.6. Synthesis of Calcium Carbonate from Yellow Phosphorus Slag

Chen et al. (2020) used yellow phosphorus slag as a raw material to obtain high-purity calcium carbonate whiskers at low temperature and pressure. During the experiments, the effect of reaction conditions, i.e., temperature, time, concentration of Ca^{2+}, ammonia dosage, and CO_2 flow rate was systematically discussed by researchers. The above mentioned parameters have an important effect on the crystal shape and microscopic morphology of the product. It was found that the content of calcium carbonate (aragonite) in the product was about 93.67% under optimal conditions. The whiteness of the product was 97.6%. The single particle diameter was found to be about 1.5–3 μm, and the length of a single particle was about 8–40 μm. SiO_2 was produced as by-products during the whole preparation process could also be reused [86]. Researches indicated that the production strategy had a good application prospect.

7. Characterization of Calcium Carbonate and Oxide Particles

The characterization of calcium carbonate NPs (CCNPs) can be done by all the sophisticated instruments but especially by FTIR, XRD, scanning electron microscopy-electron diffraction spectroscopy (SEM-EDS), TEM, and Raman spectroscopy. The characterization of CCNPs plays an important role in the confirmation of synthesis and identification of polymorphs. Fourier transform infrared spectroscopy plays an important role in the identification of functional groups in the synthesized CCNPs. A typical FTIR spectrum of a CCNPs is shown in Figure 10. The FTIR analysis of $CaCO_3$ reveals that there is a strong peak for carbonate ions due to vibration peaks $\upsilon 1$, symmetric stretching; $\upsilon 2$, out-of-plane bending; $\upsilon 3$, doubly degenerate planar asymmetric stretching; and $\upsilon 4$, doubly degenerate planar bending. The spectrum shows vibrational bands at 1456.3, 876.7, 712.7, and 409.8 cm^{-1} indicates plane bending vibration of carbonate [87]. The FTIR analysis confirmed that the $CaCO_3$ nanopowder had the characteristic peak of the carbonate group. A sharp peak at 876.7 cm^{-1} confirmed that the $CaCO_3$ nanopowder obtained from the cockleshell was calcite.

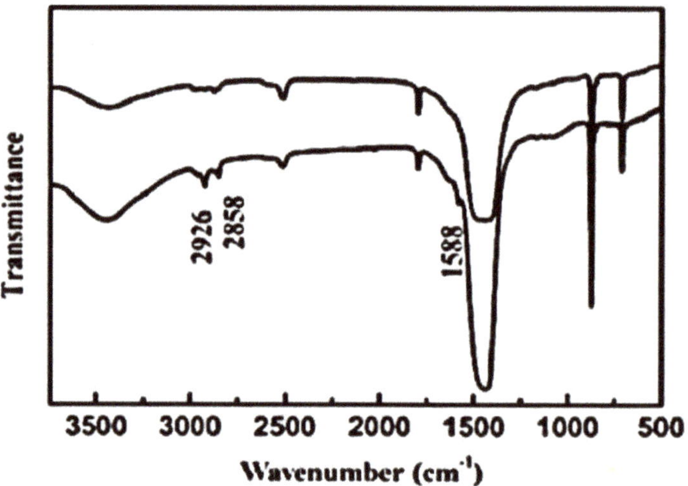

Figure 10. Fourier transform infrared spectra of calcium carbonate nanoparticles adapted from Chen et al., 2010.

Similarly, Raman spectroscopy also helps in the identification of molecules and functional group identification. The most important technique is the SEM which helps in the differentiation among the three polymorphs of CCNPs otherwise all these three polymorphs will look alike. The TEM also helps in the size determination of the particles. Based on the external morphology these polymorphs can be easily recognized as nano calcite. They are rhombohedral in shape, aragonite is needle-shaped and the vaterite is either spherical or flower shape. The typical field emission scanning electron microscope images of nano calcite, aragonite, and vaterite are shown in Figure 11A–D. The rhombohedral shaped particles confirm the calcite phases whose surface is smooth with sharp edges and corners [88].

Figure 11B is showing the aragonite phase, which is rod-shaped. The vaterite phase in Figure 11C, D is showing flower and spherical shape. The EDS attached with SEM helps in the confirmation of the formation of $CaCO_3$ along with its purity. The EDS also reveals the purity of the CCNPs by giving a molecular percentage of elements mainly C, O, and Ca in the elemental spectra. While the TEM image in Figure 12A reveals a rhombohedral shaped micron particle whose morphology is similar to the SEM images in Figure 12A [88].

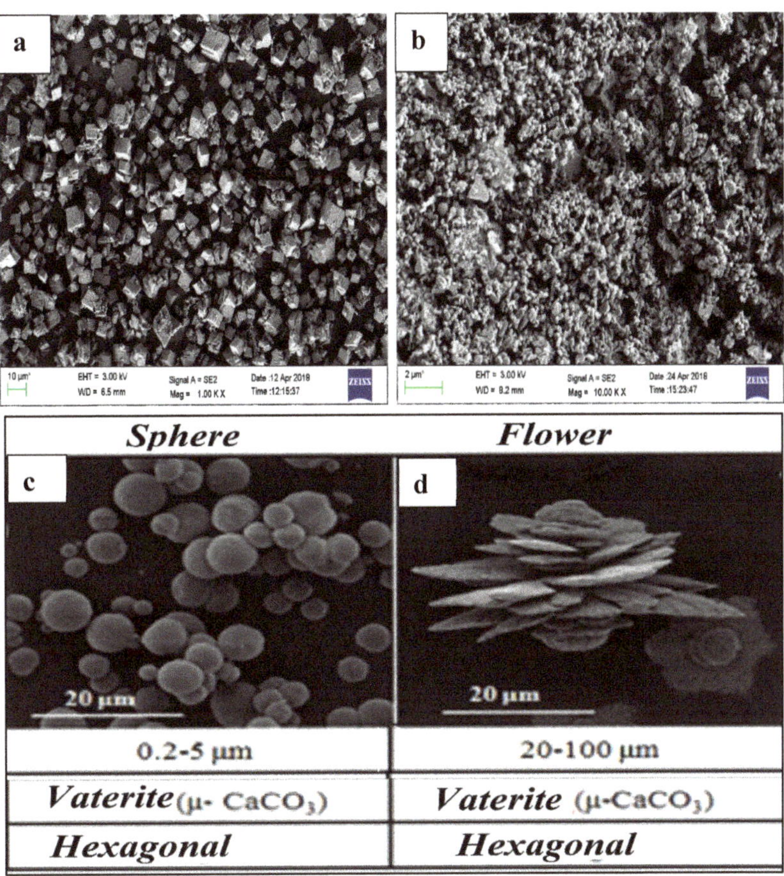

Figure 11. FESEM of (**A**) nano calcite; (**B**) aragonite; (**C**) spherical vaterite; (**D**) floral vaterite [89].

Figure 12. Transmission electron microscopic images of (**A**) calcium carbonate of microparticles and (**B**) aragonite nanorods adapted from (Islam et al., 2012).

While Figure 12B is showing nanorods whose diameter in width is below 50 nm and the length is above 200 nm. The rod-shaped particles indicate the aragonite phase of Ca NPs [88].

X-ray diffraction (XRD) analysis helps in the confirmation of mineralogy and phase determination of the CCNPs. A typical XRD pattern is shown in Figure 13, which exhibits characteristics peaks of aragonite at 2θ values of 26.34°, 33.24°, 45.98°, 33.24°, 45.98°, and 26.3°, which correlate with (hkl) indices of (111), (012), (221), and (021), (0.12), and (221), respectively [23].

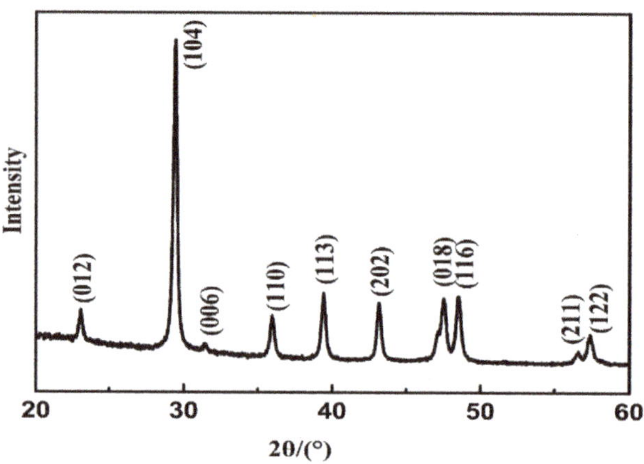

Figure 13. X-ray powder diffraction (XRD) spectrum of calcium carbonate nanoparticles adapted from Chen et al., 2010 (Chen et al., 2010).

8. Applications of Calcium Carbonate Nanoparticles

$CaCO_3$ is one of the materials which are used very frequently in the fields of ceramics [23], medicine, drug delivery, paints, fertilizers, food, cosmetics and drugs, papers, fillers, inks [9], catalysts, rubbers, plastics, and paints. In most of these industries, it is used as a viscosity modifier. It is also used as a nano adsorbent for the remediation of pollutants from the wastewater.

8.1. Applications of Calcium Carbonate Particles for Environmental Cleanup

The application of CCPs in the field of environmental cleanup is very significant due to the low cost of the nanoparticle, easy availability, and biocompatible nature [23]. It can be used for the removal of both organic as well as organic pollutants from the wastewater. Moreover, the insolubility of $CaCO_3$ in water makes them more suitable for application in aqueous solutions [90]. Among inorganic pollutants, it can be applied for the removal of heavy metals [91] from the wastewater [92], while the organic pollutants that can be removed by CCPs are phenols, pesticides, and dyes [93]. In addition, it can also be used as a nano biosensor for the detection and removal of pollutants (microbial, chemical). All these examples are given below where CCPs have been used for the remediation of inorganic and organic pollutants.

8.2. Heavy Metals Removal

Heavy metals contamination is one of the serious nuisances in the area of water and wastewater management. The elimination of heavy metals from the aqueous solutions is carried out by numerous methods that we have already seen above in the introduction section. We know that adsorption is a reliable, economical, and efficient technique [94]. Some of the most common adsorbents are zeolites, activated carbon, alumina, zinc oxide, etc. Here the size of adsorbents falls in the size of microns, so the process is less efficient.

When adsorption is done employing biological material like microbes then the process is called bioremediation [95]. When biological material and NPs are used together with either by surface modification or in the form of composite, the process is called nano bioremediation. When the adsorption process is carried out by NPs then the process is called nano remediation [96]. In the above section, we have already seen the advantages of CCPs as an adsorbent. So, here we have highlighted the examples where CCPs have been applied for the removal of heavy metals and other pollutants.

Hong et al., 2011, reported the removal of heavy metals (Cu^{2+}, Cr^{6+}, Cd^{2+}, and Pb^{2+}) by using calcium carbonate particles extracted from the starfish by using enzymes like alpha-amylase, beta-amylase, and protease. It was found that the obtained adsorbent were comparatively effective in the heavy metal removal than the calcium carbonate obtained from crabs, oyster, and cockle shells. Moreover, the calcium carbonate was highly porous in nature, which again made them suitable and effective candidate for the remediation of heavy metals [91].

The remediation of heavy metals from the aqueous solution by using CCPs synthesized through chemical method metals [97]. In this experiment, $CaCO_3$ was used to remove the Fe and Pb ions from the aqueous solutions and studied the adsorption mechanism including the kinetic and isotherm models at room temperature. From the above study following findings were obtained: (i) adsorption kinetics follows a pseudo-second-order equation; (ii) Langmuir isotherm revealed that the maximum monolayer adsorption capacity of $CaCO_3$ for Pb (II) was 1210 ± 30 mg/g and for Fe (II) ions, it was 845 ± 8 mg/g; (iii) optimal dose of $CaCO_3$ was 200 mg/L at 25 °C, and (iv) and the removal efficiency of Fe and Pb ions is enhanced by a precipitation transformation mechanism instead of adsorption.

Jacob et al. (2018) reported the use of $CaCO_3$ coated bacterial magnetosomes (biogenic magnetite obtained from the internal structures of magnetotactic bacteria) for remediation of Cr (III) and Ni (II) ions from tannery effluent. The following findings were observed from the experiment.

(i) Equilibrium was attained within an hour for Cr (III) ions at pH-6.0 and Ni (II) ions at a pH-8.0.
(ii) The adsorption process followed a pseudo-second-order reaction kinetics, along with Langmuir and Freundlich adsorption isotherms.
(iii) Adsorption of metallic ions on calcite microcrystals was spontaneous and endothermic.
(iv) Almost equal removal efficiency (Cr-94% and Ni-84%) was observed with both magnetic calcite and calcite crystals but it was higher than magnetosomes and activated carbon.
(v) Removal of both the metal ions was facilitated by applying an external magnetic field.
(vi) It is concluded that the magnetic calcite could act as a potential and alternative adsorbent for the elimination of heavy metals from tannery effluents [98].

Abeykoon et al. (2017) reported the elimination of the excess of fluoride ions from different water sources from Sri Lanka using porous CCNPs. The reported approach is quite simple, cost-effective, and innovative for the elimination of fluoride ions from the wastewater. The synthesized porous CCNPs removes the fluoride by simple adsorption mechanism and about 100% removal efficiency was achieved within one hour only [99].

8.3. As a Biosensor

CCPs can be used in developing a simple and economical biosensor due to their unique properties like high porosity, high surface area, and low mass transport barrier. Earlier investigators have reported the utilization of CCPs to detect phenols and glucose by immobilizing enzymes polyphenol oxidase (PPO) and glucose oxidase, respectively. The PPO/nano-$CaCO_3$ electrode possesses a wide detection range (i.e., 6×10^{-9} to 2×10^{-5} M), sub-nanomolar detection limit (0.44 nM at a signal to noise ratio of 3), good stability (70% remained after 56 days), the shorter response time (i.e., <12 s), large current density, and very high sensitivity (i.e., 474 mA/M). Similarly, the glucose biosensor also provided the same benefits and was very little influenced by other compounds (like ascorbic acid, glutathione, L-cysteine, and p-acetaminophenol) at their normal levels. The development of a highly sensitive amperometric phenol biosensor by using CCNPs of size around 80 nm. The sensor was developed by casting (nano-$CaCO_3$)-PPO bio composite on the surface of a glassy carbon electrode by glutaraldehyde cross-linking. The special three-dimensional structure, porous morphology, hydrophilicity, and biocompatibility of the nano-$CaCO_3$ matrix provided high enzyme loading, and the enzyme entrapped in this matrix retained its activity to a large extent. In addition, the effects of pH value applied potential, temperature, and electrode construction were also studied [100]. In one of the experiments, investigators have used the polyelectrolyte capsules (with or without $CaCO_3$ core) as a pH indicator that works even in the presence of salt in the solution. High removal capacities of heavy metals, i.e., Cd^{2+} 515, Pb^{2+} 1028, Cr^{3+} 259, Fe^{3+} 321, and Ni^{2+} 537 mg/g were achieved from aqueous solutions using precipitated amorphous calcium carbonate (ACC) NPs. The removal process was via precipitation, so the ACC became non-reusable. Few investigators have developed multi-structured $CaCO_3$/magnetite composite crystals by microwave radiation method which was used for the photocatalytic splitting of water to liberate oxygen. One special advantage with such photocatalyst is that it can be recycled after extraction from the solution.

8.4. Phenolic Compounds Removal

Phenolic compounds are a class of polluting chemicals, which are exploited extensively and discharged into the environment. These phenolic compounds have a high tendency to be absorbed by living beings from their skin and mucous membranes. Therefore, it is of utmost importance to determine the content of phenolic compounds due to their toxicity and persistent nature in the environment. a biosensor is one of the most important tools for the detection of such phenolic compounds, out of which amperometric biosensor based on PPO or tyrosinase is most widely used for the detection of phenols due to its effectiveness and simplicity [101]. PPO is a metalloenzyme that contains a binuclear copper active site and catalysis, in the presence of dioxygen, the hydroxylation of monophenols to catechols (monooxygenase activity), which in turn are oxidized to ortho-quinone (catecholase activity). The phenol biosensor transduction is thus based on the amperometric detection of the enzymatically generated o-quinone.

8.5. Dye Removal

Mosleh (2018) synthesized 15–50 nm, cobalt tungstate, and $CaCO_3$ nanocomposite for the removal of organic pollutants (dyes) from the wastewater. The morphological and chemical properties were analyzed by vibrating sample magnetometer, UV–Vis diffuse reflectance spectroscopy, XRD, and SEM where the latter two instruments helped in the confirmation of the average crystallite size of nanocomposites, whose size was in the range 15–50 nm. The high absorption with 3.2 eV bandgap of synthesized nanocomposite was identified with the DRS spectrum. The synthesized nanocomposite exhibits photocatalytic activity which was used for the degradation of methyl violet, methylene blue, phenol red, and eosin Y dyes. The developed nanocomposite degraded the MV dye up to 99%, within 200 min [102]. Ma et al. reported the synthesis of CCNPs from finger citron residue (FCR) that have already been discussed above in the synthesis section and utilized them for the removal of Congo red (CR) from the wastewater [82]. Here Nickel-doped porous $CaCO_3$ monoliths were developed from the FCR and applied for the removal of anionic dye CR by adsorption method in a batch experiment. The adsorption of CR by such NPs showed a pseudo-second-order kinetic model and Langmuir adsorption isotherm. It was found that due to the differences in positive and negative charge effects between CR and Nickel, a higher amount of CR dye was adsorbed on the Nickel-doped porous $CaCO_3$ monoliths which authenticate that later are a promising adsorbent for the removal of the anionic dyes from wastewater.

The removal of various organic and inorganic pollutants from different water and wastewater indicates that the main process for the removal of pollutants was adsorption. In some cases, CCPs were without surface functionalized while in some cases the CCPs surface were modified by various capping agents. These capping agents were used in order to make the CCPs more specific for a specific pollutants. The adsorption is a simple technique which is practiced from a century to remediate the pollutants. In this process inorganic pollutants like heavy metals gets reduced on the surface of the CCPs. Due to this reduction, the toxic heavy metals gets converted into nontoxic form which is beneficial for the flora and fauna. Moreover, adsorption process do not involve any sophisticated instruments, and machinery so it is considered easy, and economical. The only problem faced is the disposal or recycling of the CCPs loaded with the toxic heavy metals.

9. Conclusions

The CCPs have enormous potential in the field of electronics, medicine, drug delivery, paints, feed supplements, and environmental cleanup due to their biocompatibility, biodegradability, and low cost. However, the recovery of CCPs makes all these properties more justified, which further lowers the cost of the product and minimizes the pollution on the environment. Undoubtedly, the recovery of CCPs from the waste by products like fly ash, incense sticks ash, egg shell, cockle shell, gypsum, and dolomite provides an alternative for the calcium-based industries. The CCPs exhibit three polymorphisms, viz., calcite, aragonite, and vaterite. These polymorphs vary in the stability of the particle. The CCPs have the potential for the remediation of heavy metals, pesticides, dyes, and other pollutants from the wastewater. Moreover, the biodegradable, biocompatible, and economical nature of CCPs make the process of adsorption much economical.

Author Contributions: Conceptualization, V.K.Y., N.C.; methodology, V.K.Y., K.K.Y. and S.P.; validation, N.C., G.G., M.M.S.C.-P. and A.H.K.; formal analysis, V.K.Y., K.K.Y., N.C. and G.G.; resources, N.A.K., S.I., M.M.S.C.-P. and A.H.K.; writing—original draft preparation, V.K.Y.; writing—review and editing, V.K.Y., S.P., V.T. and N.C.; supervision, V.K.Y., A.H.K., V.T. and S.P.; project administration V.K.Y., K.K.Y., S.I., N.A.K. and V.T. Funding acquisition, A.H.K., S.I., N.A.K. and M.M.S.C.-P.; Software's, V.T., S.P., M.M.S.C.-P., N.C., S.I., N.A.K. and A.H.K. All authors have read and agreed to the published version of the manuscript.

Funding: The authors thankfully acknowledge the Deanship of Scientific Research, King Khalid University, Abha, KSA, for funding the project under the grant number R.G.P.1/201/41.

Institutional Review Board Statement: Not applicable.

Informed Consent Statement: Not applicable.

Data Availability Statement: Not applicable.

Acknowledgments: The authors thankfully acknowledge the Deanship of Scientific Research, King Khalid University, Abha, KSA, for funding the project under the grant number R.G.P.1/201/41.

Conflicts of Interest: The authors declare no conflict of interest.

References

1. Wagh, R.; Dongre, A. Agricultural Sector: Status, Challenges and it's Role in Indian Economy. *J. Commer. Manag. Thought* **2016**, *7*, 9. [CrossRef]
2. Khan, M.; Khan, A.U.; Hasan, M.A.; Yadav, K.K.; Pinto, M.M.C.; Malik, N.; Yadav, V.K.; Khan, A.H.; Islam, S.; Sharma, G.K. Agro-Nanotechnology as an Emerging Field: A Novel Sustainable Approach for Improving Plant Growth by Reducing Biotic Stress. *Appl. Sci.* **2021**, *11*, 2282. [CrossRef]
3. Campos, D.A.; Gómez-García, R.; Vilas-Boas, A.A.; Madureira, A.R.; Pintado, M.M. Management of Fruit Industrial By-Products-A Case Study on Circular Economy Approach. *Molecules* **2020**, *25*, 320. [CrossRef]
4. Chand Malav, L.; Yadav, K.K.; Gupta, N.; Kumar, S.; Sharma, G.K.; Krishnan, S.; Rezania, S.; Kamyab, H.; Pham, Q.B.; Yadav, S.; et al. A review on municipal solid waste as a renewable source for waste-to-energy project in India: Current practices, challenges, and future opportunities. *J. Clean. Prod.* **2020**, *277*. [CrossRef]
5. Quina, M.J.; Pinheiro, C.T. Inorganic Waste Generated in Kraft Pulp Mills: The Transition from Landfill to Industrial Applications. *Appl. Sci.* **2020**, *10*, 2317. [CrossRef]
6. Tang, P.; Chen, W.; Xuan, D.; Cheng, H.; Poon, C.S.; Tsang, D.C.W. Immobilization of hazardous municipal solid waste incineration fly ash by novel alternative binders derived from cementitious waste. *J. Hazard. Mater.* **2020**, *393*, 122386. [CrossRef]

7.   Sheets, J.L.; Wee, A.G.; Simetich, B.; Beatty, M.W. Effect of Water Dilution on Full-Arch Gypsum Implant Master Casts. *Prosthesis* **2020**, *2*, 266–276. [CrossRef]
8.   Erdogan, N.; Eken, H.A. Precipitated calcium carbonate production, synthesis and properties. *Phys. Probl. Min. Process.* **2017**, *53*, 57–68. [CrossRef]
9.   Bicchieri, M.; Valentini, F.; Calcaterra, A.; Talamo, M. Newly Developed Nano-Calcium Carbonate and Nano-Calcium Propanoate for the Deacidification of Library and Archival Materials. *J. Anal. Methods Chem.* **2017**, *2017*, 1–8. [CrossRef]
10.   Xu, Y.; Ye, J.; Zhou, D.; Su, L. Research progress on applications of calcium derived from marine organisms. *Sci. Rep.* **2020**, *10*, 18425. [CrossRef]
11.   Al Omari, M.M.H.; Rashid, I.S.; Qinna, N.A.; Jaber, A.M.; Badwan, A.A. Chapter Two—Calcium Carbonate. In *Profiles of Drug Substances, Excipients and Related Methodology*; Brittain, H.G., Ed.; Academic Press: Cambridge, MA, USA, 2016; Volume 41, pp. 31–132.
12.   Hussein, A.I.; Ab-Ghani, Z.; Che Mat, A.N.; Ab Ghani, N.A.; Husein, A.; Ab Rahman, I. Synthesis and Characterization of Spherical Calcium Carbonate Nanoparticles Derived from Cockle Shells. *Appl. Sci.* **2020**, *10*, 7170. [CrossRef]
13.   Mydin, R.B.S.; Zahidi, I.N.M.; Ishak, N.N.; Shaida, N.; Ghazali, S.N.; Moshawih, S.; Siddiquee, S. Potential of Calcium Carbonate Nanoparticles for Therapeutic Applications. *Malays. J. Med. Health Sci.* **2018**, *14*, 2636–9346.
14.   Bewernitz, M.A.; Lovett, A.C.; Gower, L.B. Liquid–Solid Core-Shell Microcapsules of Calcium Carbonate Coated Emulsions and Liposomes. *Appl. Sci.* **2020**, *10*, 8551. [CrossRef]
15.   Palacios, S.; Ramirez, M.; Lilue, M. Clinical study of the tolerability of calcium carbonate-casein microcapsules as a dietary supplement in a group of postmenopausal women. *Drugs Context* **2020**, *9*, 2020-1-4. [CrossRef]
16.   Rötzer, N.; Schmidt, M. Historical, Current, and Future Energy Demand from Global Copper Production and Its Impact on Climate Change. *Resources* **2020**, *9*, 44. [CrossRef]
17.   Jannah, Z.; Mubarok, H.; Syamsiyah, F.; H Putri, A.A.; Rohmawati, L. Preparation of Calcium Carbonate (from Shell-fish)/Magnesium Oxide Composites as an Antibacterial Agent. *Iop Conf. Ser. Mater. Sci. Eng.* **2018**, *367*, 012005. [CrossRef]
18.   Islam, K.N.; Bakar, M.Z.B.A.; Ali, M.E.; Hussein, M.Z.B.; Noordin, M.M.; Loqman, M.Y.; Miah, G.; Wahid, H.; Hashim, U. A novel method for the synthesis of calcium carbonate (aragonite) nanoparticles from cockle shells. *Powder Technol.* **2013**, *235*, 70–75. [CrossRef]
19.   Tangboriboon, N.; Kunanuruksapong, R.; Sirivat, A. Preparation and properties of calcium oxide from eggshells via calcination. *Mater. Sci. Pol.* **2012**, *30*, 313–322. [CrossRef]
20.   Singh, M.; Kumar, S.V.; Waghmare, S.; Sabale, P.D. Aragonite-vaterite-calcite: Polymorphs of CaCO3 in 7th century, C.E lime plasters of Alampur group of Temples, India. *Constr. Build. Mater.* **2016**, *112*, 386–387. [CrossRef]
21.   Habte, L.; Khan, M.D.; Shiferaw, N.; Farooq, A.; Lee, M.-h.; Jung, S.-h.; Ahn, J.W. Synthesis, Characterization and Mechanism Study of Green Aragonite Crystals from Waste Biomaterials as Calcium Supplement. *Sustainability* **2020**, *12*, 5062. [CrossRef]
22.   Trushina, D.B.; Bukreeva, T.V.; Kovalchuk, M.V.; Antipina, M.N. CaCO$_3$ vaterite microparticles for biomedical and personal care applications. *Mater. Sci. Eng. C* **2014**, *45*, 644–658. [CrossRef]
23.   Nurul Islam, K.; Abu Bakar, M.Z.; Ali, M.; Hussein, M.; Noordin, M.M.; Yusof, L.; Haron, A.W.; Hakim, M.; Bee Abd Hamid, S. Facile Synthesis of Calcium Carbonate Nanoparticles from Cockle Shells. *J. Nanomater.* **2012**, *2012*, 5. [CrossRef]
24.   Cabral Pinto, M.M.S.; Silva, M.M.V.G.; Neiva, A.M.R. Geochemistry of U-bearing minerals from the Vale de Abrutiga uranium mine area, Central Portugal. *Neues Jahrb. Für Mineral. Abh.* **2008**, *185*, 183–198. [CrossRef]
25.   Cabral Pinto, M.M.S.; Silva, M.M.V.G.; Neiva, A.M.R.; Guimarães, F.; Silva, P.B. Release, Migration, Sorption, and (Re)Precipitation of U during Peraluminous Granite Alteration under Oxidizing Conditions in Central Portugal. *Geosciences* **2018**, *8*, 95. [CrossRef]
26.   Cabral Pinto, M.; Silva, M.; Neiva, A.; Guimarães, F.; Silva, P. Uranium minerals from a Portuguese Variscan Peraluminous granite, its alteration and related uranium-quartz veins. In *Uranium: Compounds, Isotopes and Applications*; Wolfe, G., Ed.; Nova Science Publishers, Inc.: Hauppauge, NY, USA, 2009; pp. 287–318.
27.   Idrees, H.; Zaidi, S.Z.J.; Sabir, A.; Khan, R.U.; Zhang, X.; Hassan, S.U. A Review of Biodegradable Natural Polymer-Based Nanoparticles for Drug Delivery Applications. *Nanomaterials* **2020**, *10*, 1970. [CrossRef] [PubMed]
28.   Luo, X.; Song, X.; Cao, Y.; Song, L.; Bu, X. Investigation of calcium carbonate synthesized by steamed ammonia liquid waste without use of additives. *Rsc Adv.* **2020**, *10*, 7976–7986. [CrossRef]
29.   Ranjan, R.; Narnaware, S.D.; Patil, N.V. A Novel Technique for Synthesis of Calcium Carbonate Nanoparticles. *Natl. Acad. Sci. Lett.* **2018**, *41*, 403–406. [CrossRef]
30.   El-sherbiny, S.; El-Sheikh, S.; Barhoum, A. Preparation and modification of nano calcium carbonate filler from waste marble dust and commercial limestone for papermaking wet end application. *Powder Technol.* **2015**, *279*, 290–300. [CrossRef]
31.   de Beer, M.; Maree, J.P.; Liebenberg, L.; Doucet, F. Conversion of calcium sulphide to calcium carbonate during the process of recovery of elemental sulphur from gypsum waste. *Waste Manag.* **2014**, *34*, 2373–2381. [CrossRef]
32.   Yoada, R.M.; Chirawurah, D.; Adongo, P.B. Domestic waste disposal practice and perceptions of private sector waste management in urban Accra. *Bmc Public Health* **2014**, *14*, 697. [CrossRef]
33.   Yadav, V.K.; Yadav, K.K.; Gnanamoorthy, G.; Choudhary, N.; Khan, S.H.; Gupta, N.; Kamyab, H.; Bach, Q.-V. A novel synthesis and characterization of polyhedral shaped amorphous iron oxide nanoparticles from incense sticks ash waste. *Environ. Technol. Innov.* **2020**, *20*, 101089. [CrossRef]

34. Yadav, V.K.; Choudhary, N.; Heena Khan, S.; Khayal, A.; Ravi, R.K.; Kumar, P.; Modi, S.; Gnanamoorthy, G. Incense and Incense Sticks: Types, Components, Origin and Their Religious Beliefs and Importance among Different Religions. *J. Bio Innov.* **2020**, *9*, 1420–1439. [CrossRef]
35. Lin, T.-C.; Krishnaswamy, G.; S Chi, D. Incense smoke: Clinical, structural and molecular effects on airway disease. *Clin. Mol. Allergy Cma* **2008**, *6*. [CrossRef]
36. Yadav, V.K.; Singh, B.; Choudhary, N. Characterization of Indian Incense Stick Powders for their Physical, Chemical and Mineralogical Properties. *World J. Environ. Biosci.* **2020**, *9*, 39–43.
37. Abdel-Shafy, H.; Mohamed-Mansour, M. A review on polycyclic aromatic hydrocarbons: Source, environmental impact, effect on human health and remediation. *Egypt. J. Pet.* **2016**, *25*, 107–123. [CrossRef]
38. Yadav, V.K.; Gnanamoorthy, G.; Cabral-Pinto, M.M.S.; Alam, J.; Ahamed, M.; Gupta, N.; Singh, B.; Choudhary, N.; Inwati, G.K.; Yadav, K.K. Variations and similarities in structural, chemical, and elemental properties on the ashes derived from the coal due to their combustion in open and controlled manner. *Environ. Sci. Pollut. Res.* **2021**. [CrossRef] [PubMed]
39. Oral, Ç.M.; Ercan, B. Influence of pH on morphology, size and polymorph of room temperature synthesized calcium carbonate particles. *Powder Technol.* **2018**, *339*, 781–788. [CrossRef]
40. Brandelli, A.; Sala, L.; Kalil, S.J. Microbial enzymes for bioconversion of poultry waste into added-value products. *Food Res. Int.* **2015**, *73*, 3–12. [CrossRef]
41. Faridi, H.; Arabhosseini, A. Application of eggshell wastes as valuable and utilizable products: A review. *Res. Agric. Eng.* **2018**, *64*, 104–114. [CrossRef]
42. M King'ori, A. A Review of the Uses of Poultry Eggshells and Shell Membranes. *Int. J. Poult. Sci.* **2011**, *10*, 908–912. [CrossRef]
43. Zaheer, K. An Updated Review on Chicken Eggs: Production, Consumption, Management Aspects and Nutritional Benefits to Human Health. *Food Nutr. Sci.* **2015**, *06*, 1208–1220. [CrossRef]
44. Nagamalli, H.; Sitaraman, M.; Kandalai, K.K.; Mudhole, G.R. Chicken egg shell as a potential substrate for production of alkaline protease by Bacillus altitudinis GVC11 and its applications. *3 Biotech* **2017**, *7*, 185. [CrossRef] [PubMed]
45. Toro, P.; Abarca, R.; Yazdani-Pedram, M.; Luis Arias, J. Eggshell, a new bio-filler for polypropylene composites. *Mater. Lett.* **2007**, *61*, 4347–4350. [CrossRef]
46. Padhi, M.K. Importance of Indigenous Breeds of Chicken for Rural Economy and Their Improvements for Higher Production Performance. *Scientifica* **2016**, *2016*, 9. [CrossRef] [PubMed]
47. Glatz, P.; Zhihong, M.; Belinda, R. Handling and Treatment of Poultry Hatchery Waste: A Review. *Sustainability* **2011**, *3*, 216–237. [CrossRef]
48. Hassan, T.; Rangari, V.; Rana, R.; Jeelani, S. Sonochemical effect on size reduction of $CaCO_3$ nanoparticles derived from waste eggshells. *Ultrason. Sonochemistry* **2013**, *20*, 1308–1315. [CrossRef]
49. Hassan, T.; Rangari, V.; Jeelani, S. Value-Added Biopolymer Nanocomposites from Waste Eggshell-Based $CaCO_3$ Nanoparticles as Fillers. *Acs Sustain. Chem. Eng.* **2014**, *2*, 706–717. [CrossRef]
50. Hariharan, M.; Varghese, N.; Cherian, A.B.; Paul, J.; Antony, K.A. Synthesis and Characterisation of CaCO3 (Calcite) Nano particles from cockle shells Using Chitosan as Precursor. *Int. J. Sci. Res. Publ.* **2014**, *4*, 5.
51. Muhammad Mailafiya, M.; Abubakar, K.; Danmaigoro, A.; Musa Chiroma, S.; Bin Abdul Rahim, E.; Aris Mohd Moklas, M.; Abu Bakar Zakaria, Z. Cockle Shell-Derived Calcium Carbonate (Aragonite) Nanoparticles: A Dynamite to Nanomedicine. *Appl. Sci.* **2019**, *9*, 2897. [CrossRef]
52. Render, D.; Samuel, T.; King, H.; Vig, M.; Jeelani, S.; Babu, R.J.; Rangari, V. Biomaterial-Derived Calcium Carbonate Nanoparticles for Enteric Drug Delivery. *J. Nanomater.* **2016**, *2016*, 1–8. [CrossRef] [PubMed]
53. Pandita, P.; Fulekar, M.H. Egg shell waste as heterogeneous nanocatalyst for biodiesel production: Optimized by response surface methodology. *J. Environ. Manag.* **2017**, *198*, 319–329. [CrossRef]
54. Ahmad, S.; Chaudhary, S.; Pathak, V.V.; Kothari, R.; Tyagi, V.V. Optimization of direct transesterification of Chlorella pyrenoidosa catalyzed by waste egg shell based heterogenous nano—CaO catalyst. *Renew. Energy* **2020**, *160*, 86–97. [CrossRef]
55. Menon, V.; Gopakumar, K. Shellfish: Nutritive Value, Health Benefits, and Consumer Safety. *Compr. Rev. Food Sci. Food Saf.* **2017**, *16*, 1219–1242. [CrossRef]
56. Tamburini, E.; Turolla, E.; Fano, E.A.; Castaldelli, G. Sustainability of Mussel (Mytilus Galloprovincialis) Farming in the Po River Delta, Northern Italy, Based on a Life Cycle Assessment Approach. *Sustainability* **2020**, *12*, 3814. [CrossRef]
57. Hamester, M.R.R.; Balzer, P.S.; Becker, D. Characterization of calcium carbonate obtained from oyster and mussel shells and incorporation in polypropylene. *Mater. Res.* **2012**, *15*, 204–208. [CrossRef]
58. Joseph, A.M.; Snellings, R.; Van den Heede, P.; Matthys, S.; De Belie, N. The Use of Municipal Solid Waste Incineration Ash in Various Building Materials: A Belgian Point of View. *Materials* **2018**, *11*, 141. [CrossRef]
59. Cabral Pinto, M.M.S.; Ferreira da Silva, E.A. Heavy Metals of Santiago Island (Cape Verde) Alluvial Deposits: Baseline Value Maps and Human Health Risk Assessment. *Int. J. Environ. Res. Public Health* **2019**, *16*, 2. [CrossRef] [PubMed]
60. Cabral-Pinto, M.M.S.; Inácio, M.; Neves, O.; Almeida, A.A.; Pinto, E.; Oliveiros, B.; Ferreira da Silva, E.A. Human Health Risk Assessment Due to Agricultural Activities and Crop Consumption in the Surroundings of an Industrial Area. *Expo. Health* **2020**, *12*, 629–640. [CrossRef]
61. Gurbuz, A.; Sari, Y.; Yuksekdag, Z.; Cinar, B. Cementation in a matrix of loose sandy soil using biological treatment method. *Afr. J. Biotechnol.* **2011**, *10*, 7432–7440.

62. Ferronato, N.; Torretta, V. Waste Mismanagement in Developing Countries: A Review of Global Issues. *Int. J. Environ. Res. Public Health* **2019**, *16*, 1060. [CrossRef]
63. Cabral Pinto, M.M.S.; Marinho-Reis, P.; Almeida, A.; Pinto, E.; Neves, O.; Inácio, M.; Gerardo, B.; Freitas, S.; Simões, M.R.; Dinis, P.A.; et al. Links between Cognitive Status and Trace Element Levels in Hair for an Environmentally Exposed Population: A Case Study in the Surroundings of the Estarreja Industrial Area. *Int. J. Environ. Res. Public Health* **2019**, *16*, 4560. [CrossRef]
64. Puyol, D.; Batstone, D.J.; Hülsen, T.; Astals, S.; Peces, M.; Krömer, J.O. Resource Recovery from Wastewater by Biological Technologies: Opportunities, Challenges, and Prospects. *Front. Microbiol.* **2017**, *7*, 2106. [CrossRef]
65. Maree, J.P.; Zvinowanda, C.M.; Mujuru, M.; Matsapola, R.M.; Delport, D.J.; Louw, a.W.J. Recovery of Calcium Carbonate from Wastewater Treatment Sludge Using a Flotation Technique. *Chem. Eng. Process Technol.* **2012**, *3*, 6. [CrossRef]
66. Xu, J.; Yan, C.; Zhang, F.; Konishi, H.; Xu, H.; Teng, H.H. Testing the cation-hydration effect on the crystallization of Ca-Mg-CO3 systems. *Proc. Natl. Acad. Sci. USA* **2013**, *110*, 17750–17755. [CrossRef] [PubMed]
67. Wonyen, D.; Kromah, V.; Gibson, B.; Nah, S.; Chelgani, S. A Review of Flotation Separation of Mg Carbonates (Dolomite and Magnesite). *Minerals* **2018**, *8*, 354. [CrossRef]
68. Yildirim, M.; Akarsu, H. Preparation of magnesium oxide (MgO) from dolomite by leach-precipitation-pyrohydrolysis process. *Phys. Probl. Min. Process.* **2009**, *44*, 15.
69. Somarathne, Y.R.; Mantilaka, P.; Karunaratne, D.G.G.P.; Rajapakse, R.; Pitawala, H.M.T.G.; Wijayantha, K. Synthesis of high purity calcium carbonate micro- and nano-structures on polyethylene glycol templates using dolomite. *Cryst. Res. Technol.* **2016**, *51*, 207–214. [CrossRef]
70. Mulopo, L.; Radebe, V. Recovery of calcium carbonate from waste gypsum and utilization for remediation of acid mine drainage from coal mines. *Water Sci. Technol.* **2012**, *66*, 1296–1300. [CrossRef] [PubMed]
71. Sultan, Z.; Sheikh, Z.; Zafar, M.S.; Sauro, S. *Dental Materials (Principles and Applications)*; Paramount Book Publishers: Karachi, Pakistan, 2018.
72. Ko, J.H.; Xu, Q.; Jang, Y.-C. Emissions and Control of Hydrogen Sulfide at Landfills: A Review. *Crit. Rev. Environ. Sci. Technol.* **2015**, *45*, 2043–2083. [CrossRef]
73. Xu, Q.; Townsend, T.; Bitton, G. Inhibition of hydrogen sulfide generation from disposed gypsum drywall using chemical inhibitors. *J. Hazard. Mater.* **2011**, *191*, 204–211. [CrossRef]
74. Sufang, W.; Lan, P. Method for Preparing a Nano-Calcium Carbonate Slurry from Waste Gypsum as Calcium Source, the Product and Use thereof. US8846562B2, 31 October 2013.
75. Huber, M.; Stark, W.; Loher, S.; Maciejewski, M.; Krumeich, F.; Baiker, A. Flame synthesis of calcium carbonate nanoparticles. *Chem. Commun.* **2005**, 648–650. [CrossRef] [PubMed]
76. Oyetunji, A.; Umunakwe, R.; Omotayo Adewuyi, B.; Samuel Nwigwe, U.; Janefrances Umunakwe, I. Evaluating the properties of nanoparticles of calcium carbonate obtained from the shells of african giant land snails (Achatina achatina) via in situ deposition technique. *Upb Sci. Bull. Ser. B: Chem. Mater. Sci.* **2019**, *81*, 86–94.
77. Chilakala, R.; Thenepalli, T.; Huh, J.-H.; Ahn, J.-W. Precipitated Calcium Carbonate Synthesis by Simultaneous Injection to Produce Nano Whisker Aragonite. *J. Korean Ceram. Soc.* **2016**, *53*, 222–226. [CrossRef]
78. Chilakala, R.; Thenepalli, T.; Huh, J.-H.; Ahn, J.-W. Preparation of Needle like Aragonite Precipitated Calcium Carbonate (PCC) from Dolomite by Carbonation Method. *J. Korean Ceram. Soc.* **2016**, *53*, 7–12. [CrossRef]
79. Zema, D.A.; Calabrò, P.S.; Folino, A.; Tamburino, V.; Zappia, G.; Zimbone, S.M. Valorisation of citrus processing waste: A review. *Waste Manag.* **2018**, *80*, 252–273. [CrossRef]
80. Mahato, N.; Sinha, M.; Sharma, K.; Koteswararao, R.; Cho, M.H. Modern Extraction and Purification Techniques for Obtaining High Purity Food-Grade Bioactive Compounds and Value-Added Co-Products from Citrus Wastes. *Foods* **2019**, *8*, 523. [CrossRef] [PubMed]
81. Khan, S.A.; Ahmad, R.; Asad, S.; Muhammad, S. Citrus flavonoids: Their biosynthesis, functions and genetic improvement. In *Citrus Molecular Phylogeny, Antioxidant Properties and Medicinal Uses*, 1st ed.; Hayat, K., Ed.; Nova Science Publishers: New York, NY, USA, 2014; pp. 31–51.
82. Kasana, R.C.; Panwar, N.R.; Kaul, R.K.; Kumar, P. Biosynthesis and effects of copper nanoparticles on plants. *Environ. Chem. Lett.* **2017**, *15*, 233–240. [CrossRef]
83. Gong, R.; Ye, J.; Dai, W.; Yan, X.; Hu, J.; Hu, X.; Li, S.; Huang, H. Adsorptive Removal of Methyl Orange and Methylene Blue from Aqueous Solution with Finger-Citron-Residue-Based Activated Carbon. *Ind. Eng. Chem. Res.* **2013**, *52*, 14297–14303. [CrossRef]
84. Mihara, N.; Soya, K.; Kuchar, D.; Fukuta, T.; Matsuda, H. Utilization of calcium sulfide derived from waste gypsum board for metal-containing wastewater treatment. *Glob. Nest J.* **2008**, *10*, 101–107.
85. Brooks, M.W.; Lynn, S. Recovery of Calcium Carbonate and Hydrogen Sulfide from Waste Calcium Sulfide. *Ind. Eng. Chem. Res. Ind Eng Chem Res* **1997**, *36*, 4236–4242. [CrossRef]
86. Chen, Q.; Ding, W.; Peng, T.; Sun, H. Synthesis and characterization of calcium carbonate whisker from yellow phosphorus slag. *Open Chem.* **2020**, *18*, 347–356. [CrossRef]
87. Abdolmohammadi, S.; Siyamak, S.; Ibrahim, N.A.; Yunus, W.M.Z.W.; Rahman, M.Z.A.; Azizi, S.; Fatehi, A. Enhancement of Mechanical and Thermal Properties of Polycaprolactone/Chitosan Blend by Calcium Carbonate Nanoparticles. *Int. J. Mol. Sci.* **2012**, *13*, 4508–4522. [CrossRef] [PubMed]

88. Wan, C.; Wang, L.-T.; Sha, J.-Y.; Ge, H.-H. Effect of Carbon Nanoparticles on the Crystallization of Calcium Carbonate in Aqueous Solution. *Nanomaterials* **2019**, *9*, 179. [CrossRef]
89. Yadav, V.K.; Choudhary, N.; Khan, S.H.; Malik, P.; Inwati, G.K.; Suriyaprabha, R.; Ravi, R.K. Synthesis and Characterisation of Nano-Biosorbents and Their Applications for Waste Water Treatment. In *Handbook of Research on Emerging Developments and Environmental Impacts of Ecological Chemistry*; Gheorghe Duca, A.V., Ed.; IGI Global: Hershey, PA, USA, 2020; pp. 252–290. [CrossRef]
90. Han, S.-J.; Yoo, M.; Kim, D.-W.; Wee, J.-H. Carbon Dioxide Capture Using Calcium Hydroxide Aqueous Solution as the Absorbent. *Energy Fuels* **2011**, *25*, 3825–3834. [CrossRef]
91. Hong, K.-S.; Myoung Lee, H.; Seong Bae, J.; Gyu Ha, M.; Sung Jin, J.; Hong, T.E.; Pil Kim, J.; Jeong, E. Removal of Heavy Metal Ions by using Calcium Carbonate Extracted from Starfish Treated by Protease and Amylase. *J. Anal. Sci. Technol.* **2011**, *2*, 75–82. [CrossRef]
92. Park, H.; Wook Jeong, S.; Yang, J.-K.; Gil Kim, B.; Lee, S.-M. Removal of Heavy Metals Using Waste Eggshell. *J. Environ. Sci. (China)* **2007**, *19*, 1436–1441. [CrossRef]
93. Liu, Y.; Jiang, Y.; Hu, M.; Li, S.; Zhai, Q.-G. Removal of triphenylmethane dyes by calcium carbonate–lentinan hierarchical mesoporous hybrid materials. *Chem. Eng. J.* **2015**, *273*, 371–380. [CrossRef]
94. Yadav, V.K.; Fulekar, M.H. Biogenic synthesis of maghemite nanoparticles (γ-Fe2O3) using Tridax leaf extract and its application for removal of fly ash heavy metals (Pb, Cd). *Mater. Today Proc.* **2018**, *5*, 20704–20710. [CrossRef]
95. Sales da Silva, I.G.; Gomes de Almeida, F.C.; Padilha da Rocha e Silva, N.M.; Casazza, A.A.; Converti, A.; Asfora Sarubbo, L. Soil Bioremediation: Overview of Technologies and Trends. *Energies* **2020**, *13*, 4664. [CrossRef]
96. Singh, R.; Behera, M.; Kumar, S. Nano-bioremediation: An Innovative Remediation Technology for Treatment and Management of Contaminated Sites. In *Bioremediation of Industrial Waste for Environmental Safety*; Bharagava, R., Ed.; Springer: Singapore, 2020; pp. 165–182. [CrossRef]
97. Mohammadifard, H.; Amiri, M.C. On tailored synthesis of nano CaCO3 particles in a colloidal gas aphron system and evaluating their performance with response surface methodology for heavy metals removal from aqueous solutions. *J. Water Environ. Nanotechnol.* **2018**, *3*, 141–149. [CrossRef]
98. Jacob, J.; Varalakshmi, R.; Gargi, S.A.; Jayasri, M.; Suthindhiran, K. Removal of Cr (III) and Ni (II) from tannery effluent using calcium carbonate coated bacterial magnetosomes. *Npj Clean Water* **2018**, *1*. [CrossRef]
99. Abeykoon, K.G.M.D.; Dunuweera, S.P.; Rajapakse, R.M.G. Synthesis of porous cakcium carbonate nanoparticles and isotherm studies for the removal of flouride in different water sources as a solution for CKD. In Proceedings of the 4th International Conference on Nanoscience and Nanotechnology 2017(ICNSNT 2017), Colombo, Sri Lanka, 14–15 December 2017; p. 1.
100. Shan, D.; Wang, Y.; Xue, H.; Cosnier, S. Sensitive and selective xanthin amperometric sensors based on calcium carbonate nanoparticles. *Sens. Actuators B Chem.* **2009**, *136*, 510–515. [CrossRef]
101. Gul, I.; Ahmad, M.S.; Naqvi, S.S.; Hussain, A.; Wali, R.; Farooqi, A.A.; Ahmed, I. Polyphenol oxidase (PPO) based biosensors for detection of phenolic compounds: A Review. *J. Appl. Biol. Biotechnol.* **2017**, *5*, 13. [CrossRef]
102. Mosleh, M. Application of new method for the synthesis of cobalt tungstate nanostructures and cobalt tungstate/calcium carbonate nanocomposites and removal of organic pollutants. *J. Mater. Sci. Mater. Electron.* **2018**, *29*, 4855–4861. [CrossRef]

Review

Agro-Nanotechnology as an Emerging Field: A Novel Sustainable Approach for Improving Plant Growth by Reducing Biotic Stress

Masudulla Khan [1,*], Azhar U. Khan [2,*], Mohd Abul Hasan [3], Krishna Kumar Yadav [4], Marina M. C. Pinto [5,*], Nazia Malik [6], Virendra Kumar Yadav [7,*], Afzal Husain Khan [8], Saiful Islam [3] and Gulshan Kumar Sharma [9]

1 Department of Botany, Aligarh Muslim University, Aligarh U.P. 202002, India
2 School of Life and Basic Sciences, Department of Chemistry, SIILAS CAMPUS, Jaipur National University, Jaipur 302017, India
3 Civil Engineering Department, College of Engineering, King Khalid University, Abha 62529, Saudi Arabia; mohad@kku.edu.sa (M.A.H.); sfakrul@kku.edu.sa (S.I.)
4 Institute of Environment and Development Studies, Bundelkhand University, Kanpur Road, Jhansi 284128, India; envirokrishna@gmail.com
5 Geobiotec Research Centre, Department of Geoscience, University of Aveiro, 3810-193 Aveiro, Portugal
6 Department of Chemistry, Aligarh Muslim University, Aligarh U.P. 202002, India; nazia.amu123@rediffmail.com
7 School of Lifesciences, SIILAS CAMPUS, Jaipur National University, Jaipur 302017, India
8 Civil Engineering Department, College of Engineering, Jazan University, Jazan 114, Saudi Arabia; ahkhan@jazanu.edu.sa
9 National Bureau of Soil Survey and Land Use Planning, Regional Centre Jorhat, Assam 785001, India; gulshansharma2222@gmail.com
* Correspondence: masudkhann@gmail.com (M.K.); azhar.u.kh@gmail.com (A.U.K.); marinacp@ua.pt (M.M.C.P.); virendra.yadav@jnujaipur.ac.in (V.K.Y.)

Citation: Khan, M.; Khan, A.U.; Hasan, M.A.; Yadav, K.K.; Pinto, M.M.C.; Malik, N.; Yadav, V.K.; Khan, A.H.; Islam, S.; Sharma, G.K. Agro-Nanotechnology as an Emerging Field: A Novel Sustainable Approach for Improving Plant Growth by Reducing Biotic Stress. *Appl. Sci.* **2021**, *11*, 2282. https://doi.org/10.3390/app11052282

Academic Editor: Anthony William Coleman

Received: 27 January 2021
Accepted: 23 February 2021
Published: 4 March 2021

Publisher's Note: MDPI stays neutral with regard to jurisdictional claims in published maps and institutional affiliations.

Abstract: In the present era, the global need for food is increasing rapidly; nanomaterials are a useful tool for improving crop production and yield. The application of nanomaterials can improve plant growth parameters. Biotic stress is induced by many microbes in crops and causes disease and high yield loss. Every year, approximately 20–40% of crop yield is lost due to plant diseases caused by various pests and pathogens. Current plant disease or biotic stress management mainly relies on toxic fungicides and pesticides that are potentially harmful to the environment. Nanotechnology emerged as an alternative for the sustainable and eco-friendly management of biotic stress induced by pests and pathogens on crops. In this review article, we assess the role and impact of different nanoparticles in plant disease management, and this review explores the direction in which nanoparticles can be utilized for improving plant growth and crop yield.

Keywords: plant diseases; nanoparticles; diseases; biotic stress; management; silver nanoparticles; zinc nanoparticles

1. Introduction

Crop cultivators suffer from high yield loss caused by various diseases. Biotic stress induced by microbes on crop plants reduces the crop yield and decreases the quality. Biotic stress causes disease in crops, which leads to the suffering of the plant. Diseases of the plant need to be controlled to maintain the abundance of food produced by farmers around the world. The management of crop diseases is very necessary to fulfill the food demand. Potato blight disease caused by plant pathogenic fungus *Phytopthora* caused more than one million deaths in Ireland [1]. Around 20–40% of agricultural crop yield losses occur globally due to various diseases caused by phytopathogenic bacteria, phytopathogenic fungi, pests, and weeds [2].

It is estimated that in 2050 the world's human population will reach around 10 billion, and around 800 million people in the world will be hungry and around 653 million people in the world will be undernourished in 2030, thus fulfilling the food demand will remain a huge challenge. The current research progress and disease management strategies are not enough to fulfill the food demand by 2050 [3]. The first green revolution made a huge difference in yield and food production, but in the last few years' crop production has been stagnant and food demand is increasing sharply, so now we need a second green revolution to fulfill the food demand of the population.

Different approaches are used by farmers to mitigate the impact of plant diseases. The agriculture system mainly relies on chemicals to manage crop diseases and inhibit the growth of phytopathogens, which cause diseases before and after crop harvesting. The excessive use of chemical pesticides, herbicides, and fungicides that are mainly used to control plant diseases causes harmful environmental and human health consequences. Tilman et al. [4] observed that the high use of chemical pesticides increases resistance in pathogens and pests, reduces nitrogen fixation, and the bioaccumulation of toxic pesticides occurs.

An example is the synthetic chemical pesticide DDT, dichlorodiphenyltrichloroethane, which was extensively used in agriculture for controlling plant pathogens and was found to be genotoxic in humans, causing endocrine disorders [5]. Water and soil pollution is also caused by the excessive use and misuse of these chemicals. There is an increasing demand day by day to reduce the use of synthetic chemicals. Consequently, the harmful effects of chemicals on wildlife, the environment, and human health have increased the need for alternative measures in the control of plant pathogens, so that some phytopathologists have focused their research on developing a new alternative that should replace the use of chemicals in controlling plant diseases.

Nanotechnology has revolutionized agriculture and can control plant diseases, although the field of nanotechnology is still in the nascent stage and needs more research analysis [6].The use of nanomaterials in agriculture will reduce the excessive use of toxic chemicals used for plant disease management (Figures 1 and 2).

"Nano" denotes one-billionth part, thus nanotechnology deals with small things. The word nano is used for materials with a size range of 0.1 to 100 nanometers [7,8]. The first time the term nanotechnology was used was by Taniguchi in 1974 to the science that largely deals with particles of nano size (1.0×10^{-9} m). When a bulk material is reduced to nano size, it has a high surface-to-volume ratio that may increase its reactivity and express some new properties [7,9]. The control of plant diseases and improving plant growth by the use of nanomaterials are some of the possible key applications in the area of plant pathology. Approximately 260,000–309,000 metric tons of nanoparticles were produced in 2010 globally, and the worldwide consumption of nanomaterials was approximately from 225,060 metric tons to 585,000 metric tons in 2014 to 2019 [10,11].

In this review article, recent research progress and the application of various nanoparticles for the sustainable management of the biotic stress of crop systems and impact on plant growth have been discussed. We try to cover the various problems associated with crop cultivation and plant diseases and the use of different nanomaterials to control phytopathogens and improve plant growth.

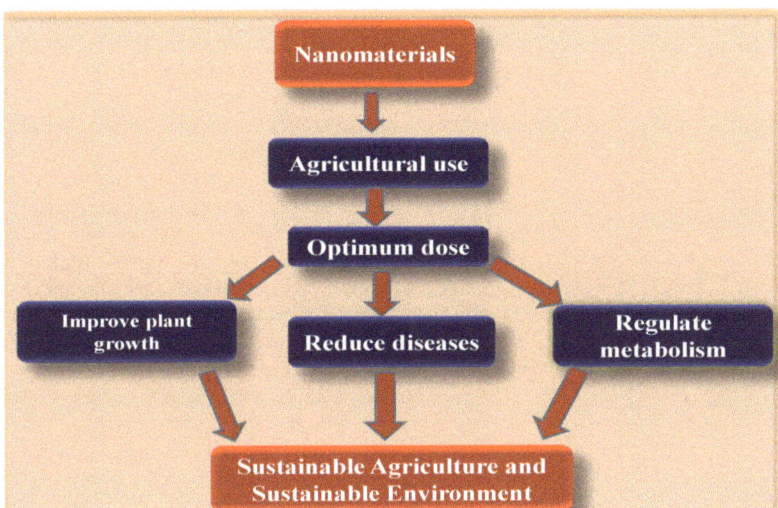

Figure 1. Schematic presentation of nanomaterials in agriculture [12].

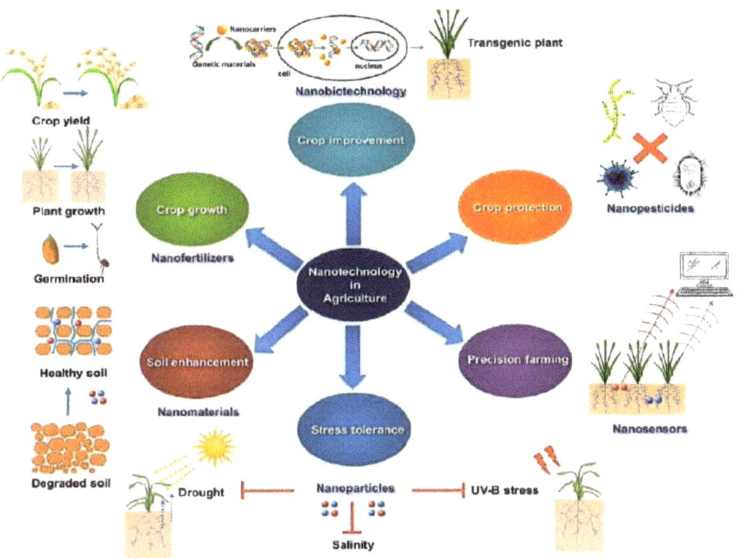

Figure 2. Various applications of nanotechnology in agriculture taken from [12].

2. Nanomaterials in Improving Plant Growth and Yield

Currently, around 1300 nanomaterials, with widespread potential applications, are available [13,14]. Nanoparticles can penetrate the cell wall because the cell wall is porous to 3.5–20 nm macromolecules. Nanoparticles can enter through stomatal openings. When stomata are present at the lower surface of leaves, the entry of nanoparticles (NPs) becomes difficult [15]. It is reported that nanoparticles of size ≤43 nm can penetrate and enter into stomata [16,17].

The effect of nanoparticles on crop plants is concentration-based. Many plant processes such as seed germination and plant growth are affected by NP concentration [18]. Many NPs have been reported to be beneficial for plant growth. Mahmoud et al. [19] used

Zn, B, Si, zeolite NPs on a potato plant and found that these nanoparticles have a positive effect on potato plants and they improve the plant growth. Khan and Siddiqui [20] treated eggplant with ZnONPs and found a foliar spray of ZnONPs causes the highest improvement in eggplant growth. Awasthi et al. [21] reported that ZnONPs have a positive effect on seed germination in the *Triticum aestivum* plant. Zinc oxide nanoparticles (ZnONPs) can enhance plant biomass and agriculture production [22]. Sabir et al. [23] also showed that nanocalcite ($CaCO_3$) application with Fe_2O_3, nano SiO_2, and MgO improved the uptake of Mg, Ca, and Fe, and also notably enhanced the intake of P with micronutrients Zn and Mn. Venkatachalam et al. [24] found that ZnONPs increase in photosynthetic pigment in the *Leucaena leucocephala* plant. Narendhran et al. [25] reported high chlorophyll-a', chlorophyll-'b' and total chlorophyll content in the *Sesamum indicum* plant when treated with ZnO NPs. Taheri et al. [26] observed that treatment of ZnONPs increases the increase in shoot dry matter in *Zea mays*. Tarafdar et al. [27] found that ZnONPs enhanced shoot and grain yield in the *Pennisetum glaucum* plant.

The application of titanium dioxide (TiO_2) on crops promotes plant growth parameters and can enhance the photosynthetic rate. Siddiqui et al. [28] used TiO_2 and ZnONPs on beet root plants. They found that both NPs increased chlorophyll and carotenoid content, improved plant growth, and also improved super oxide dismutase (SOD), catalase (CAT), H_2O_2, and proline content in plants. ZnONPs were found to be better than TiO_2NPs on beetroot plants. Raliya et al. [29] reported that TiO_2NPs treatment improved shoots in the *Vigna radiate* plant. Lawre and Raskar [30] observed that TiO_2NPs at a lower concentration enhanced seed germination and seedling growth in onion plants. Rafique et al. [31] found a positive effect of TiO_2NPs on the *Triticum aestivum* plant. Mahmoodzadeh et al. [32] found a positive effect of TiO_2NPs on the seed germination of the *Brassica napus* plant. Qi et al. [33] reported that treatment of TiO_2NPs promotes photosynthetic rate in tomato plants.

Silicon is an important element that plays a key role in several metabolic and physiological activities in plants [34]. SiO_2nanoparticles have the potential to enhance the germination and seedling growth of *Agropyron elongatum* [35]. Nano-SiO_2 can be used to produce effective fertilizers for crops and to minimize the loss of fertilizer through slow and controlled release, allowing for regulated, responsive, and timely delivery [36]. Siddiqui et al. [37] found improved seed germination in the *Cucurbita pepo* plant after treatment with Nano SiO_2. Haghighi and Pessarakli [38] reported that Nano Si treatment on the tomato plant improves photosynthetic rate in treated plants.

Copper is an essential element for plant growth and development. Copper plays a key role in the activity of many plant enzymes. Copper nanoparticles (Cu NP) are used as antimicrobial agents, gas sensors, catalysts, electronics, etc. [39]. Wang et al. [40] found that CuO NPs improved photosynthesis in the *Spinacia oleracea* plant. Zhao et al. [41] reported that $Cu(OH)_2$NPs improved the antioxidant system of the *Lactuca sativa* plant. Shinde et al. [42] found that $Mg(OH)_2$NP treatment promotes seed germination and seedling growth in the *Zea mays* plant. Hussain et al. [43] reported that MgO NPs improve the antioxidant system in *Raphanus sativus* plants. Cai et al. [44] observed that MgO NPs can promote the plant growth of the Tobacco plant. Imada et al. [45] found that MgO NPs can induce resistance in the tomato plant.

Iqbal et al. [46] reported that AgNP treatment improved plant growth and tolerance to heat stress in the *Triticum aestivum* plant. Mehta et al. [47] found that AgNPs' foliar application enhanced growth and biomass in the *Vigna sinensis* plant. Pilon et al. [48] observed that chitosan NPs protect apple plants after post-harvest. Van et al. [49] found that chitosan NPs improve plant growth in Robusta coffee.

Das et al. [50] found that FeS_2 NPs improved seed germination in *Cicer arietinum*, *Daucus carota*, *pinacia oleracea*, *Brassica juncea*, and *Sesamum indicum* crops. The effects of various nanomaterials have been summarized in the following table (Table 1).

Table 1. Effect of various nanomaterials on plant physiology and growth parameters.

Nanoparticles	Plant	Effect on Plants in a Dose-Dependent Manner	Reference
Zn, B, Si, Zeolite NPs	Potato	Improve plant growth	[19]
ZnO NPs	Eggplant	Increase plant growth attributes	[20]
ZnO NPs	*Triticum aestivum*	Positive effect on seed germination	[21]
SiO$_2$ & TiO$_2$ NPs	Rice	Improve plant growth attributes	[22]
Nano-size calcite product [CaCO3(40%), SiO2(4%), MgO (1%), and Fe2O3(1%)]	Grapevine	Increase plant growth attributes and photosynthetic pigment	[23]
ZnO NPs	*Leucaena leucocephala*	Increase in photosynthetic pigment and total soluble protein contents	[24]
ZnO NPs	*Sesamum indicum*	High chlorophyll'a', chlorophyll'b', and total chlorophyll content level	[25]
ZnO NPs	*Zea mays*	Increased shoot dry matter and leaf area indexes.	[26]
ZnO NPs	*Pennisetum glaucum*	ZnO NPs enhanced shoot and grain yield	[27]
TiO$_2$ & ZnO NPs	Beetroot	Increased plant growth and shoot dry matter	[28]
TiO$_2$ NPs	*Vigna radiata* L.	Improvement was observed in shoot length	[29]
TiO$_2$ NPs	Onion	Lower concentration of TiO$_2$ NPs enhanced seed germination and seedlings growth	[30]
TiO$_2$ NPs	*Triticum aestivum* L.	Increase in the plant's root and shoot lengths	[31]
TiO$_2$ NPs	*Brassica napus*	Promoted seed germination and seedling vigor improved	[32]
TiO$_2$ NPs	Tomato	Promote the photosynthetic rate	[33]
SiO$_2$NPs	*Larix olgensis*	Increase in plant height, root length, and chlorophyll content	[34]
SiO$_2$NPs	*Agropyron elongatum* L.	Improve seed germination	[35]
Nano- SiO$_2$	*Cucurbita pepo* L.	Reduce the salt stress effect	[37]
Nano Si	Tomato	Enhancement of germination rate and dry weight	[38]
CuO NPs	Spinacia oleracea	Improved photosynthesis in treated plants	[40]
MgO NPs	Tobacco	Promote plant growth	[44]
MgO NPs	Tomato	Induce resistance in tomato plant	[45]
AgNPs	Wheat	Regulate antioxidative defence system	[46]
AgNPs	soil bacterial diversity	Regulate soil bacterial diversity	[47]
Chitosan NPs	Apples	They reduce microbial growth	[48]
Chitosan NPs	Robusta cofee	Improved growth parameters	[49]
FeS$_2$ NPs	*Cicer arietinum; pinacia oleracea; Daucus carota, Brassica juncea and Sesamum indicum*	Seed germination enhanced in tested crops	[50]
Chitosan NPs	Rice	Reduces disease severity	[51]
Chitosan NPs	Strawberry	Regulate defense response	[52]
SiNPs	*Helianthus annuus*	Improved germination	[53]
SilicaNPs	*Vicia faba* L.	Improved growth parameters	[54]
SiO$_2$NPs	Pea	Improved growth parameters and chlorophyll content	[55]
SiO$_2$ & MoNPs	Rice	Regulate seed germination	[56]
SiO$_2$NPs	*Indocalamus barbatus*	Improved photosynthetic pigments	[57]

Table 1. *Cont.*

Nanoparticles	Plant	Effect on Plants in a Dose-Dependent Manner	Reference
SilicaNPs	*Zea mays. L*	Improve silica content in plants	[58]
SiO_2NPs	Maize	Improved growth parameters and increased seed stability	[59]
$SiO_2$and TiO_2NPs	Soybean	Enhance germination of seeds	[60]
$Cu(OH)_2$	*Lactuca sativa*	Improve antioxidant system	[61]
$Cu(OH)_2$	Spinach	Improve the antioxidant system	[62]
ZnO NPs	*Glycine max*	Enhanced Antioxidant system	[63]
ZnO NPs	Cabbage, cauliflower, and tomato	Enhance pigments, protein, and sugar contents	[64]
ZnO NPs	*Arachis hypogaea*	Seed germination enhaced	[65]
FeS_2 NP	Spinach	Improve plant growth	[66]
TiO_2 NPs	*Glycine max L.*	Positive effect on the seed and oil yield and component compared to the control	[67]
TiO_2 NPs	*Mentha Piperita*	Increased root length	[68]
TiO_2 NPs	*Agropyron desertorum*	Improves seed germination	[69]

3. Nanomaterials in Various Diseases Management

Nanomaterials have antimicrobial activity. Silver nanoparticles have anti-bacterial and anti-fungal properties. Kim et al. [70] have reported the fungicidal effects of nano-silver against *Alternaria alternata, A. brassicicola, A. solani, Botrytis cinerea, Cladosporium cucumerinum, Corynespora cassiicola, Cylindrocarpon destructans, Didymella bryoniae, Fusarium oxysporum f. sp. cucumerinum, F. oxysporum f. sp. lycopersici, F. oxysporum, F. solani, Fusarium sp., Glomerella cingulata* and a few other fungi. Gautam et al. [71] showed the antifungal and antibacterial activity of AgNPs against *Erwinia sp., Bacillus megaterium, Pseudomonas syringe, Fusarium graminearum, F. avenaceum,* and *F. culmorum* fungi. Rodríguez-Serrano et al. [72] reported the antibacterial activity of AgNPs against *E. coli.* Husseinet al. [73] reported the antibacterial activity of AgNPs against *Staphylococcus aureus* and *Klebsiella pneumonia.* Shehzad et al. [74] reported that AgNPs have antibacterial activity against Gram-positive (*Bacillus subtilis*) and Gram-negative (*Escherichia coli*) bacteria. Mohanta et al. [75] reported that AgNPs have antibacterial activity against food borne pathogens *Pseudomonas aeruginosa, Escherichia coli,* and *Bacillus subtilis.* Abdelmale and Salaheldin [76] reported that AgNPs show antifungal activity against *Alternaria alternata, A. citri,* and *Penicillium digitatum* fungi. Krishnaraj et al. [77] found theantifungal activity of AgNPs against *Alternaria alternata, Macrophomina phaseolina, Botrytis cinerea, Sclerotinia sclerotiorum, Curvularia lunata,* and *Rhizoctonia solani* fungi. Jo et al. [78] described the antifungal activity of AgNPs against *Bipolaris sorokiniana* and *Magnaporthe grisea* fungi.

Shahryari et al. [79] reported that AgNPs and a silver–chitoson composite show antibacterial activity against *Pseudomonas syringae* pv. *syringae* bacteria. Divya et al. [80] reported that chitoson NPs have antifungal activity against *Macrophomia phaseolina* and *Alternaria alterneta* fungi. Xing et al. [81] reported that chitoson NPs have antifungal activity against *Fusarium solani* and *Aspergillus niger* fungi. Dang et al. [82] reported that AuNPs have antibacterial activity against *E. coli* bacteria. Attar and Yapaoz [83] observed that ZnO and AuNPs have antibacterial activity against *E. coli* bacteria. The gold nanoparticles showed toxic effect on bacteria, *Salmonella typhimurium,* in which the macro gold did not exhibit. Jayaseelana et al. [84] synthesized gold nanoparticles from *Abelmoschus esculentus* and reported their antifungal activity. The antifungal activity of AuNPs was tested against *Puccinia graministritci, Aspergillus niger, Aspergillus flavus* and *Candida albicans* using the standard well diffusion method. The maximum zone of inhibition was observed in the Au NPs against *P. graminis* and *C. albicans.*

Fan et al. [85] observed the antibacterial activity of Cu composites against *Xanthomonas euvesicatoria*. Huang et al. [86] showed the antifungal activity of CuO NPs against *Botrytis cinerea, Colletotrichum graminicola, Rhizoctonia solani, Colletotrichum musae, Magnaporthe oryzae, Penicillium digitatum,* and *Sclerotium rolfsii.* Giannousiet al. [87] showed the antifungal activity of CuO and Cu$_2$O NPs against *Phytophthora infestans.* Sharmaet al. [88] reported the antifungal and antibacterial activity of MgONPs against *Ralstonia solanacearum* bacteria and *Phomopsis vexans* fungus. Imada et al. [45] found the antibacterial activity of MgONPs against *Ralstonia solanacearum.* Derbalah et al. [89] observed the antifungal property of silica NPs against *Alternaria solani* fungus. Akpinar et al. [90] found that SiO$_2$ NPs possess antifungal properties against *Fusarium oxysporum* f. sp. *lycopersici* and *F. oxysporum* f. sp. *radicislycopersici.* Park et al. [91] showed the antifungal activity of Nano Si-Ag against *Pythium ultimum, Magnaporthe grisea, Colletotrichum gloeosporioides, Botrytis cineria, Rhizoctonia solani, Pseudomonas syringae, Xanthomonas compestris* pv. *vesicatoria.*

Jamdagni et al. [92] found that ZnO NPs have promising antifungal activity against *Alternaria alternate Botrytis cinerea, Aspergillus niger, Fusarium oxysporum,* and *Penicillium expansum* fungi. Navale et al. [93] found the promising antifungal activity of ZnO NPs against *Aspergillus flavus* and *Aspergillus fumigates* fungi. Rajiv et al. [94] reported the antifungal activity of ZnO NPs against *Aspergillus flavus, A. niger, A. fumigates, Fusarium culmorum,* and *F. oxysporium.* Gunalan et al. [95] found that ZnO NPs have promising antifungal activity against *Aspergillus flavus, Trichoderma harzianum, A. nidulans,* and *Rhizopus stolonifer.* Dimkpa et al. [96] have shown the antifungal activity of ZnO nanoparticles on *Fusarium graminearum* fungus. Jayaseelan et al. [97] synthesized ZnO nanoparticles using *Aeromonas hydrophila* and screened their activity against pathogenic bacteria *P. aeruginosa,* and fungi, *C. albicans, A. flavus,* and *A. niger.* Sar et al. [98] reported the antifungal activity of TiO$_2$ NPs against *Fusarium oxysporum* f. sp. *radicislycopersici* and *Fusarium oxysporum* f. sp. *Lycopersici.* Hamza et al. [99] found the antifungal activity of TiO$_2$ NPs against *Cercospora beticola.* Ardakani [100] found the nematicidal activity of TiO$_2$ NPs against *Meloidogyne incognita* nematode. Kasemets et al. [101] reported the antifungal activity of ZnO and TiO$_2$ NPs against *Saccharomyces cerevisiae.* Cui et al. [102] found that TiO$_2$ NPs have antibacterial against *P. syringae pv. lachrymans* and *P. cubensis* (Table 2, Figure 3).

Table 2. Various nanomaterials in plant disease management

Nanoparticle	Pathogen	Effect	Reference
Ag NPs	*Alternaria alternata, A. brassicicola, A. solani, Cladosporium cucumerinum, Botrytis cinerea, Corynespora cassiicola, Cylindrocarpon destructans, Didymella bryoniae, F. oxysporum f. sp. lycopersici, F. oxysporum, Fusarium oxysporum f.sp. cucumerinum, F. solani, Fusarium sp., Glomerella cingulata, P. spinosum, Monosporascuscannonballus, Pythium aphanidermatum, Stemphylium lycopersici*	Show antifungal activity	[70]
AgNPs	*Erwinia sp., Bacillus megaterium, Pseudomonas syringe, Fusarium graminearum, F. avenaceum, F. culmorum*	An inhibitory effect on tested microbes	[71]
AgNPs	*Escherichia coli*	Antibacterial activity	[72]
AgNPs	*Staphylococcus aureus* and *Klebsiella pneumonia*	Antibacterial activity	[73]
AgNPs	Gram-positive (*Bacillus subtilis*) and gram-negative (*Escherichia coli*).	An inhibitory effect on tested bacteria	[74]
AgNPs	Foodborne pathogens viz. *Pseudomonas aeruginosa, Escherichia coli, Bacillus subtilis.*	Antibacterial activity	[75]
AgNPs	*Alternaria alternata, A. citri, Penicillium digitatum*	Show antifungal properties	[76]
AgNPs	*Alternaria alternata, Macrophomina phaseolina, Botrytis cinerea, Sclerotinia Sclerotiorum, Curvularia lunata, Rhizoctonia solani*	Show Antifungal activity.	[77]

Table 2. *Cont.*

Nanoparticle	Pathogen	Effect	Reference
AgNPs	*Bipolaris sorokiniana* and *MagnaportheGrisea*	Show antifungal activity	[78]
AgNPs and Cs-Ag nanocomposite	*Pseudomonas syringaepv.syringae*	Show antibacterial activity	[79]
Chitosan NPs	*Klebsiella pneumoniae, Escherichia coli, Staphylococcus aureus, Pseudomonas aeruginosa*	Show antibacterial activity	[80]
Chitosan NPs	*Fusarium solani, Aspergillus niger*	Show Antifungal activity	[81]
Au NPs	*Escherichia coli* and *Staphylococcus*	Antibacterial activity	[82]
ZnO and Au NPs	*E. coli*	Antibacterial activity	[83]
AuNPs	*Puccinia graminis tritci, Aspergillus flavus, Aspergillus niger* and *Candida albicans*	Show Antifungal activity	[84]
Cu composites	*Xanthomonas euvesicatoria*	Antibacterial activity	[85]
CuO NPs	*Botrytis cinerea, Colletotrichumgraminicola, Rhizoctonia solani, Colletotrichum musae, Magnaportheoryzae, Penicillium digitatum, Sclerotium rolfsii*	Show antifungal activity	[86]
CuO and Cu$_2$O NPs	*Phytophthora infestans*	Show antifungal activity	[87]
MgO NPs	*Ralstonia solanacearum, Phomopsis vexans*	Show antifungal and antibacterial activity	[88]
SilicaNPs	*Alternaria sp*	Show antifungal activity	[89]
SiO$_2$ NPs	*Fusarium oxysporum* f. sp. *lycopersici* and *F. oxysporum* f. sp. *radicislycopersici*	Possess antifungal properties	[90]
Nano Si-Ag	*Pythium ultimum, Magnaporthe grisea, Colletotrichum gloeosporioides, Botrytis cineria, Rhizoctonia solani, Pseudomonas syringae, Xanthomonas compestris* pv. *vesicatoria*	Show antifungal and antibacterial activity	[91]
ZnO NPs	*Alternaria alternate Botrytis cinerea, Aspergillus niger, Fusarium oxysporum* and *Penicillium expansum*	Antifungal activity against all the tested fungi	[92]
ZnO NPs	*Aspergillus flavus* and *Aspergillus fumigates*	Shown potential activity against these tested fungi	[93]
ZnO NPs	*Aspergillus flavus, A. niger, A. fumigatus Fusarium culmorum* and *F. oxysporium*	The highest zone of inhibition occurred in *A. flavus*	[94]
ZnO NPs	*Aspergillus flavus, A. nidulans, Trichoderma harzianum* and *Rhizopus stolonifer*	Antifungal activity	[95]
ZnO NPs	*Fusarium graminearum*	Antifungal activity	[96]
ZnO NPs	*Pseudomonas aeruginosa*	Antibacterial activity	[97]
TiO$_2$ NPs	*Fusarium oxysporum* f. sp. *radicislycopersici* and *Fusarium oxysporum* f. sp. *Lycopersici*	Antifungal activity	[98]
TiO$_2$ NPs	*Cercosporabeticola*	Pathogen growth was inhibited	[99]
TiO$_2$ NPs	*Meloidogyne incognita*	Controlled *M. incognita*	[100]
TiO$_2$ NPs and ZnO NPs	*Saccharomyces cerevisiae*	Antifungal activity	[101]
TiO$_2$ NPs	*P. syringaepv. lachrymans* and *P. cubensis*	Reduced infection of pathogen	[102]
Metallic NPs	*Fungus and Bacteria*	Antibacterial and antifungal activity	[103]
Metallic NPs	*Microbes*	Antibacterial and antifungal activity	[104]
AgNPs	*Fusarium culmorum*	Antifungal activity	[105]

Table 2. *Cont.*

Nanoparticle	Pathogen	Effect	Reference
Chitosan NPs	*Streptococcus*	Antibacterial activity	[106]
AuNPs	*Candida albicans*	Antifungal activity	[107]
AuNPs	*Escherichia coli, Staphylococcus aureus*	Antibacterial activity	[108]
ZnO NPs	*Ralstonia solanacearum*	Antibacterial activity	[109]
ZnO NPs	*Botrytis, Escherichia*	Antibacterial and antifungal activity	[110]
ZnO NPs	*Fusarium oxysporum, Aspergillus niger*	Antibacterial and antifungal activity	[111]
ZnO NPs	*Alternaria alternate, Fusarium oxysporum, Rhizopus stolonifer* and *Mucor plumbeus*	Inhibit germination of spores of fungi	[112]
ZnO NPs	*Botrytis cinerea* and *Penicillium expansum*	Significantly inhibit growth	[113]
ZnO NPs	*Psedomanas* sp. and *Fusarium* sp.	Antibacterial and antifungal activity	[114]
TiO$_2$ NPs	*Xanthomonas hortorum* pv. *pelargonii, X. axonopodis* pv. *Poinsettiicola*	Antibacterial activity	[115]

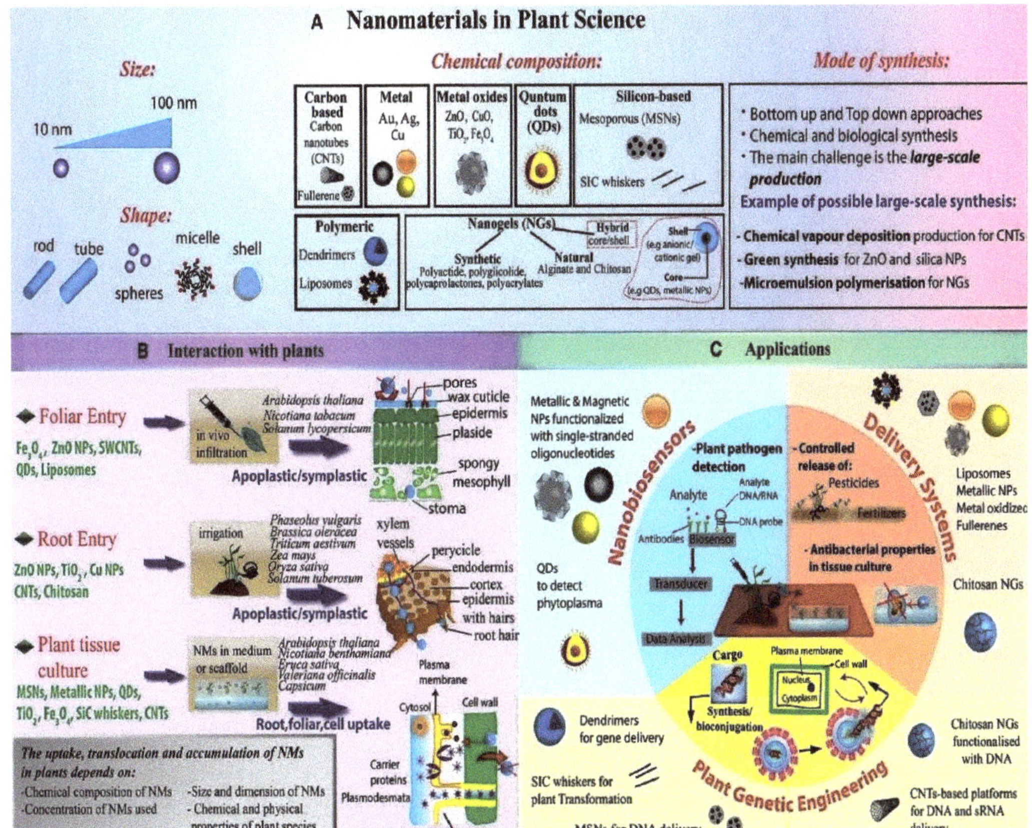

Figure 3. (**A**) Different types of nanoparticles. (**B**) Schematic presentation of delivery methods of different nanoparticles and translocation in plants. (**C**) Various applications of nanoparticles (Taken from Sanzari et al. [116]).

The inhibitory action of nanoparticles on fungi and bacteria includes disruption by pore formation in the cell membrane, disturbance in membrane potential, cell wall damage, direct attachment to the cell surface, DNA damage, cell cycle arrest, the inhibition of enzyme activity and reactive oxygen species (ROS) generation, and this finally leads to death. Nanoparticles generate the ROS, which causes damage to the cellular structures. The different components of reactive oxygen species include free radicals, such as hydrogen peroxide (H_2O_2), superoxide (O_2^-), singlet oxygen ($^1O_2$), carbon dioxide radical (CO_2^-), hydroxyl ($HO^\cdot$), hydroperoxyl (HO_2), carbonate (CO_3^-), peroxyl (RO_2), and alkoxyl (RO), and non-radicals, such as ozone (O_3), nitric oxide (NO), hypobromous acid (HOBr), hypochlorous acid (HOCl), hypochlorite (OCl^-), peroxy nitrite ($ONOO^-$), organic peroxides (ROOH), peroxo monocarbonate ($HOOCO_2^-$), peroxy nitrous acid (ONOOH) and peroxy nitrate (O_2NOO-), and these nanoparticles accumulate in the membrane of bacteria or fungi, which leads to change in the permeability of the cell membrane and disturbs the proton motive force (PMF).Oxidative stress due to the higher concentration leads to single- and double-strand breaks and nitrogen base and pentose sugar lesions [103,104].

4. Toxic Effect of Nanoparticles

Nanomaterials' effect on organisms is largely dependent on the dose, size, and shape, the types of NPs, concentration, and the duration of exposure to NPs and the plant/animal species [117,118]. Nanoparticles at optimum concentration augment the plant's growth, but high concentrations of nanoparticles could be toxic for plants. Kushwah and Patel [119] observed that the optimum concentration of nano TiO_2 in the *Vicia faba* plant ranged from 5–50 mg/L. Other studies proved that TiO_2 NPs may induce stress in plants such as tomato, cucumber and spinach at high concentration [120]. Silver nanoparticles cause chromosomal aberrations in *Vicia faba* [121]. Lopez-Moreno et al. [122] reported that CeO_2 nanoparticles can induce DNA damage in soybean.

5. Conclusions

In summary, the literature shows that food demands will increase with time, and to fulfill the demand of people, the present agricultural practices are not sufficient and chemicals used in agriculture as pesticides have a severe toxic effect on the environment. Thus, we need to develop an alternative approach that has a less toxic effect on the environment and that could help in fulfilling food demands. According to estimates, around 192.8 Mt chemical fertilizers were used in 2016–2017 in the whole world. The use of toxic chemicals and pesticides causes environmental pollution, which affects fauna and flora. Pathogens and pests induce resistance against fungicides and pesticides. Hence, optimizing of the use of toxic chemical pesticides and fungicides is needed. Nanotechnology is flamboyant and has provided nanostructure materials as pesticide and fertilizer carriers. Nanomaterials can develop smart fertilizers as they can enhance nutrient availability and reduce environmental pollution [123]. Novel nanotechnology can be an alternative that can reduce crop diseases and enhance crop yield. Previous studies reported a significant positive effect of nanomaterials on crop plants. This novel technology can reduce the use of toxic chemicals and pesticides that contaminate soil, the environment, and groundwater. Further research is needed to develop this technology on a large scale (Figure 4).

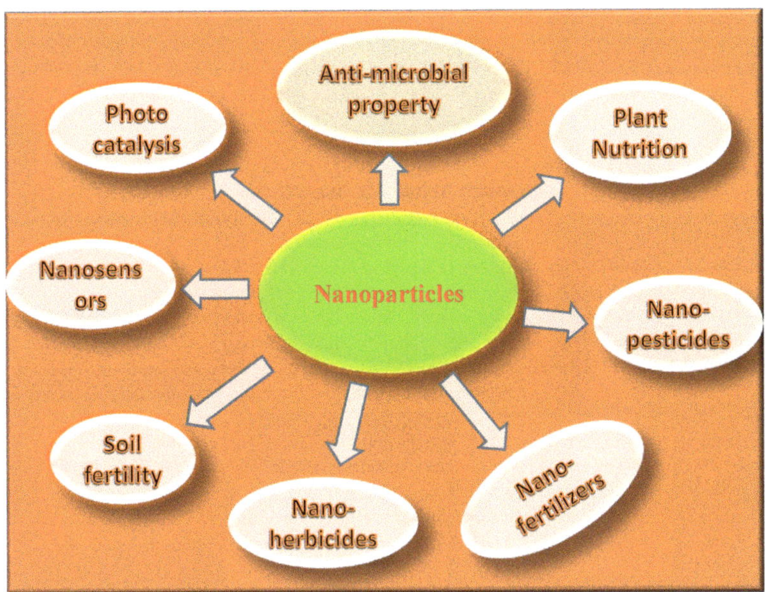

Figure 4. Diagram showing general applications of nanoparticles in agriculture.

Author Contributions: Writing—original draft: M.K. and A.U.K.; Writing—review & editing: M.K., A.U.K., V.K.Y., K.K.Y. and M.M.C.P.; Conceptualization: G.K.S. and S.I. Data curation: M.A.H. and A.H.K.; Formal analysis: K.K.Y., N.M. and G.K.S.; Funding acquisition: M.A.H., M.M.C.P., A.H.K. and S.I.; Investigation: N.M., S.I. and G.K.S.; Methodology: M.K., A.U.K. and M.M.C.P.; Project administration: M.K., A.U.K., M.M.C.P.; Resources: M.A.H., K.K.Y., S.I., G.K.S. and A.H.K.; Software: K.K.Y., G.K.S. and N.M.; Supervision: A.U.K., M.M.C.P., A.H.K., V.K.Y., Validation: V.K.Y., M.A.H. and A.H.K.; Visualization: V.K.Y., S.I. and N.M. All authors have read and agreed to the published version of the manuscript.

Funding: Funding for this work has been provided by the Deanship of Scientific Research, KKU, Abha, Kingdom of Saudi Arabia, under research grant award number R.G.P2/85/41.

Institutional Review Board Statement: Not Applicable.

Informed Consent Statement: Not Applicable.

Data Availability Statement: The raw data used for this proposed work have been cited in the manuscript. Moreover, the derived data supporting the findings of this study have been graphically depicted and are available with the corresponding author on request.

Acknowledgments: The authors thankfully acknowledge the Deanship of Scientific Research, King Khalid University, Abha, for providing administrative and financial support. Funding for this work has been provided by the Deanship of Scientific Research, KKU, Abha, Kingdom of Saudi Arabia, under research grant award number R.G.P2/85/41. The authors also acknowledges the contribution and support provided by the University of Aveiro, Portugal. The authors wish to acknowledge the work of all the references used in this study.

Conflicts of Interest: The authors declare that there is no conflict of interest regarding the publication of this paper.

References

1. O'Neill, J.R. *Irish Potato Famine*; BDO: Edina, MN, USA, 2009; p. 1604535148.
2. Cabral-Pinto, M.M.S.; Inácio, M.; Neves, O.; Almeida, A.A.; Pinto, E.; Oliveiros, B.; Ferreira da Silva, E.A. Human health risk assessment due to agricultural activities and crop consumption in the surroundings of an industrial area. *Expo. Health* **2020**, *12*, 629–640. [CrossRef]

3. F.A.O.; World Health Organization; WHO Expert Committee on Food Additives. *Evaluation of Certain Contaminants in Food: Eighty-Third Report of the Joint FAO/WHO Expert Committee on Food Additives*; World Health Organization: Geneve, Switzerland, 2017.

4. Tilman, D.; Knops, J.; Wedin, D.; Reich, P. Plant diversity and composition: Effects on productivity and nutrient dynamics of experimental grasslands. In *Biodiversity and Ecosystem Functioning*; Loreau, M., Naeem, S., Inchausti, P., Eds.; Oxford University Press: Oxford, UK, 2002; pp. 21–35.

5. Cohn, B.A.; Wolff, M.S.; Cirillo, P.M.; Sholtz, R.I. DDT and breast cancer in young women: New data on the significance of age at exposure. *Environ. Health Perspect.* **2007**, *115*, 1406–1414. [CrossRef] [PubMed]

6. Cabral-Pinto, M.M.S.; Marinho-Reis, P.; Almeida, A.; Pinto, E.; Neves, O.; Inácio, M.; Gerardo, B.; Freitas, S.; Simões, M.R.; Dinis, P.A.; et al. Links between cognitive status and trace element levels in hair for an environmentally exposed population: A case study in the surroundings of the Estarreja industrial area. *Int. J. Environ. Res. Public Health* **2019**, *16*, 4560. [CrossRef]

7. Khan, A.U.; Khan, M.; Khan, M.M. Antifungal and Antibacterial Assay by Silver Nanoparticles Synthesized from Aqueous Leaf Extract of *Trigonella foenum-graecum*. *BioNanoScience* **2019**. [CrossRef]

8. Khan, M.; Khan, A.U.; Alam, M.J.; Park, S.; Alam, M. Biosynthesis of silver nanoparticles and its application against phytopathogenic bacterium and fungus. *Int. J. Environ. Anal. Chem.* 2019. [CrossRef]

9. Cabral-Pinto, M.M.S.; Silva, E.A.F.; Silva, M.M.V.G.; Melo-Gonçalves, P.; Candeias, C. Environmental risk assessment based on high-resolution spatial maps of potentially toxic elements sampled on stream sediments of Santiago, Cape Verde. *Geosciences* **2014**, *4*, 297–315. [CrossRef]

10. Cabral Pinto, M.M.S.; Marinho-Reis, P.; Almeida, A.; Freitas, S.; Simões, M.R.; Diniz, L.; Pinto, E.; Ramos, P.; Ferreira da Silva, E.; Moreira, P.I. Fingernail trace element content in environmentally exposed individuals and its influence on their cognitive status in ageing. *Expo. Health* **2019**, *11*, 181–194. [CrossRef]

11. BBC. BCC Research, Global Markets for Nanocomposites, Nanoparticles, Nanoclays, and Nanotubes. 2014. Available online: https://www.bccresearch.com/market-research/nanotechnology/nanocomposites-market-nan021f.html?vsmaid=203/ (accessed on 19 October 2017).

12. Shang, Y.; Hasan, M.; Ahammed, G.J.; Li, M.; Yin, H.; Zhou, J. Applications of nanotechnology in plant growth and crop protection: A review. *Molecules* **2019**, *24*, 2558. [CrossRef]

13. Aslani, F.; Bagheri, S.; Muhd, J.N.; Juraimi, A.S.; Hashemi, F.S.; Baghdadi, A. Effects of engineered nanomaterials on plants growth: An overview. *Sci. World J.* **2014**. [CrossRef] [PubMed]

14. Srivastav, A.; Yadav, K.K.; Yadav, S.; Gupta, N.; Singh, J.K.; Katiyar, R.; Kumar, V. *Nano-Phytoremediation of Pollutants from Contaminated Soil Environment: Current Scenario and Future Prospects*; Springer: Berlin/Heidelberg, Germany, 2018; pp. 383–401.

15. Chichiriccò, G.; Poma, A. Penetration and toxicity of nanomaterials in higher plants. *Nanomaterials* **2015**, *5*, 851–873. [CrossRef] [PubMed]

16. Eichert, T.; Kurtz, A.; Steiner, U.; Goldbach, H.E. Size exclusion limits and lateral heterogeneity of the stomatal foliar uptake pathway for aqueous solutes and water-suspended nanoparticles. *Physiol. Plant* **2008**, *134*, 151–160. [CrossRef]

17. Schreiber, L. Polar paths of diffusion across plant cuticles: New evidence for an old hypothesis. *Ann. Bot.* **2005**, *95*, 1069–1073. [CrossRef] [PubMed]

18. Zheng, L.; Hong, F.; Lu, S.; Liu, C. Effect of nano-TiO_2 on strength of naturally aged seeds and growth of spinach. *Biol. Trace Elem. Res.* **2005**, *104*, 83–91. [CrossRef]

19. Mahmoud, A.W.M.; Abdeldaym, E.A.; Abdelaziz, S.M.; El-Sawy, M.B.I.; Mottaleb, S.A. Synergetic Effects of Zinc, Boron, Silicon, and Zeolite Nanoparticles on Confer Tolerance in Potato Plants Subjected to Salinity. *Agronomy* **2020**, *10*, 19. [CrossRef]

20. Khan, M.; Siddiqui, Z.A. Zinc oxide nanoparticles for the management of *Ralstonia solanacearum*, *Phomopsis vexans* and *Meloidogyne incognita* incited disease complex of eggplant. *Indian Phytopathol.* **2018**, *71*, 355–364. [CrossRef]

21. Awasthi, A.; Bansal, S.; Jangir, L.K.; Awasthi, G.; Awasthi, K.K.; Awasthi, K. Effect of ZnO Nanoparticles on Germination of *Triticum aestivum* Seeds. *Macromol. Symp.* **2017**, *376*, 1700043. [CrossRef]

22. Rizwan, M.; Ali, S.; Rehman, M.Z. Effect of foliar applications of silicon and titanium dioxide nanoparticles on growth, oxidative stress, and cadmium accumulation by rice (*Oryza sativa*). *Acta Physiol. Plant.* **2019**, *41*, 35–42. [CrossRef]

23. Sabir, A.; Yazar, K.; Sabir, F.; Kara, Z.; Yazici, M.A.; Goksu, N. Vine growth, yield, berry quality attributes and leaf nutrient content of grapevines as influenced by seaweed extract (*Ascophyllum nodosum*) and nanosized fertilizer pulverizations. *Sci. Hortic.* **2014**, *175*, 1–8. [CrossRef]

24. Venkatachalam, P.; Jayaraj, M.; Manikandan, R.; Geetha, N.; Rene, E.R.; Sharma, N.C.; Sahi, S.V. Zinc oxide nanoparticles (ZnONPs) alleviate heavy metal-inducedtoxicity in Leucaena leucocephala seedlings: A physiochemicalanalysis. *Plant Physiol. Biochem.* **2017**, *110*, 59–69. [CrossRef]

25. Narendhran, S.; Rajiv, P.; Rajeshwari, S. Influence of zinc oxide nanoparticles on growth of *Sesamum Indicum*, L. in zinc deficient soil. *Int. J. Pharm. Pharm. Sci.* **2016**, *8*, 365–371.

26. Taheri, M.; Qarache, H.A.; Qarache, A.A.; Yoosefi, M. The effects of zinc-oxide nanoparticles on growth parameters of corn (SC704). *STEM Fellowsh. J.* **2015**, *1*, 17–20. [CrossRef]

27. Tarafdar, J.C.; Raliya, R.; Mahawar, H.; Rathore, I. Development of zinc nanofertilizer to enhance crop production in pearl millet (*Pennisetum americanum*). *Agric.Res.* **2014**, *3*, 257–262. [CrossRef]

28. Siddiqui, Z.A.; Khan, M.R.; Abdallah, E.F.; Parveen, A. Titanium dioxide and zinc oxide nanoparticles affect some bacterial diseases, and growth and physiological changes of beetroot. *Int. J. Veg. Sci.* **2019**, *25*, 409–430. [CrossRef]

29. Raliya, R.; Biswas, P.; Tarafdar, J. TiO$_2$ nanoparticle biosynthesis and its physiological effect on mung bean (*Vigna radiata* L.). *Biotechnol. Rep.* **2015**, *5*, 22–26. [CrossRef]

30. Laware, S.L.; Raskar, S. Effect of titanium dioxide nanoparticles on hydrolytic and antioxidant enzymes during seed germination in onion. *Int. J. Curr. Microbiol. Appl. Sci.* **2014**, *3*, 749–760.

31. Rafique, R.; Arshad, M.; Khokhar, M.; Qazi, I.; Hamza, A.; Virk, N. Growth response of wheat to titania nanoparticles application. *NUST J. Eng. Sci.* **2014**, *7*, 42–46.

32. Mahmoodzadeh, H.; Nabavi, M.; Kashefi, H. Effect of Nanoscale Titanium Dioxide Particles on the Germination and Growth of Canola (*Brassica napus*). *J. Ornam. Hortic. Plant* **2013**, *3*, 25–32.

33. Qi, M.; Liu, Y.; Li, T. Nano-TiO$_{(2)}$ improve the photosynthesis of tomato leaves under mild heat stress. *Biol. Trace Elem. Res.* **2013**, *156*, 323–328. [CrossRef] [PubMed]

34. Bao, S.L.; Chun, H.L.; Li, J.F.; Shu, C.Q.; Min, Y. Effect of TMS (nanostructured silicon dioxide) on growth of Changbai larch seedlings. *J. For. Res.* **2004**, *15*, 138–140.

35. Azimi, R.; Borzelabad, M.J.; Feizi, H.; Azimi, A. Interaction of SiO$_2$ nanoparticles with seed prechilling on germination and early seedling growth of tall wheatgrass (*Agropyron elongatum* L.). *Pol. J. Chem. Technol.* **2014**, *16*, 25–29. [CrossRef]

36. Nair, R.; Varghese, S.H.; Nair, B.G.; Maekawa, T.; Yoshida, Y.; Kumar, D.S. Nanoparticulate material delivery to plants. *Plant Sci.* **2010**, *179*, 154–163. [CrossRef]

37. Siddiqui, M.H.; Al-Whaibi, M.H.; Faisal, M.; Al Sahli, A.A. Nano-silicon dioxide mitigates the adverse effects of salt stress on *Cucurbita pepo* L. *Environ. Toxicol. Chem.* **2014**, *33*, 2429–2437. [CrossRef]

38. Haghighi, M.; Afifipour, Z.; Mozafarian, M. The Effect of N-Si on Tomato Seed Germination under Salinity Levels. *J. Biol. Environ. Sci.* **2012**, *6*, 87–90.

39. Kasana, R.C.; Panwar, N.R.; Kaul, R.K.; Kumar, P. Biosynthesis and effects of copper nanoparticles on plants. *Environ. Chem. Lett.* **2017**, *15*, 233–240. [CrossRef]

40. Wang, Y.; Lin, Y.; Xu, Y.; Yin, Y.; Guo, H.; Du, W. Divergence in response of lettuce (var. ramosa Hort.) to copper oxide nanoparticles/microparticles as potential agricultural fertilizer. *Environ. Pollut. Bioavailab.* **2019**, *31*, 80–84. [CrossRef]

41. Zhao, L.J.; Peralta-Videa, J.R.; Rico, C.M.; Hernandez-Viezcas, J.A.; Sun, Y.; Niu, G.; Servin, A.; Nunez, J.E.; Duarte-Gardea, M.; Gardea-Torresdey, J.L. CeO$_2$ and ZnO nanoparticles change the nutritional quality of cucumber (*Cucumis sativus*). *J. Agric. Food Chem.* **2014**, *62*, 2752–2759. [CrossRef]

42. Shinde, S.; Paralikar, P.; Ingle, A.P.; Rai, M. Promotion of seed germination and seedling growth of Zea mays by magnesium hydroxide nanoparticles synthesized by the filtrate from Aspergillus niger. *Arab. J. Chem.* **2020**, *13*, 3172–3182. [CrossRef]

43. Hussain, F.; Hadi, F.; Akbar, F. Magnesium oxide nanoparticles and thidiazuron enhance lead phyto accumulation and antioxidative response in *Raphanus sativus* L. *Environ. Sci. Pollut. Res.* **2019**, *26*, 30333–30347. [CrossRef]

44. Cai, L.; Liu, M.; Liu, Z.; Yang, H.; Sun, X.; Chen, J.; Ding, W. MgONPs can boost plant growth: Evidence from increased seedling growth, morpho-physiological activities, and Mg uptake in tobacco (*Nicotiana tabacum* L.). *Molecules* **2018**, *23*, 3375. [CrossRef] [PubMed]

45. Imada, K.; Sakai, S.; Kajihara, H.; Tanaka, S.; Ito, S. Magnesium oxide nanoparticles induce systemic resistance in tomato against bacterial wilt disease. *Plant Pathol.* **2016**, *65*, 551–560. [CrossRef]

46. Iqbal, M.; Raja, N.I.; Wattoo, F.H.; Hussain, M.; Ejaz, M.; Saira, H. Assessment of AgNPs exposure on physiological and biochemical changes and antioxidative defence system in wheat (*Triticum aestivum* L) under heat stress. *IET NanobioTechnol.* **2018**, *16*, 230–236. [CrossRef]

47. Mehta, C.M.; Srivastava, R.; Arora, S.; Sharma, A.K. Impact assessment of silver nanoparticles on plant growth and soil bacterial diversity. *3 Biotech* **2016**, *6*, 254.

48. Pilon, L.; Spricigo, P.C.; Miranda, M.; Moura, M.R.; Assis, O.B.G.; Mattoso, L.H.C.; Ferreira, M.D. Chitosan nanoparticle coatings reduce microbial growth on fresh-cut apples while not afecting quality attributes. *Int. J. Food Sci. Technol.* **2015**, *50*, 440–448. [CrossRef]

49. Van, S.N.; Minh, H.D.; Anh, D.N. Study on chitosan nanoparticles on biophysical characteristics and growth of robusta cofee in green house. *Biocatal. Agric. Biotechnol.* **2013**, *2*, 289–294.

50. Das, C.K.; Srivastava, G.; Dubey, A.; Roy, M.; Jain, S.; Sethy, N.K.; Saxena, M.; Harke, S.; Sarkar, S.; Misra, K. Nano-iron pyrite seed dressing: A sustainable intervention to reduce fertilizer consumption in vegetable (beetroot, carrot), spice (fenugreek), fodder (alfalfa), and oilseed (mustard, sesamum) crops. *Nanotechnol. Environ. Eng.* **2016**, *1*, 2. [CrossRef]

51. Manikandan, A.; Sathiyabama, M. Preparation of chitosan nanoparticles and its effect on detached rice leaves infected with *Pyricularia grisea*. *Int. J. Biol. Macromol.* **2016**, *1*, 58–61. [CrossRef]

52. Hajirasouliha, M.; Jannesari, M.; Najafabadi, F.S.; Hashemi, M. Effect of novel chitosan nanoparticle coating on postharvest qualities of strawberry. In Proceedings of the 4th International Conference of Nanostructures, Kish Island, Iran, 12–14 March 2012.

53. Janmohammadi, M.; Sabaghnia, N. Effect of pre-sowing seed treatments with silicon nanoparticles on germinability of sunflower (*Helianthus annuus*). *Botanica* **2015**, *21*, 13–21. [CrossRef]

54. Roohizadeh, G.; Majd, A.; Arbabian, S. The effect of sodium silicate and silica nanoparticles on seed germination and growth in the *Vicia faba* L. *Trop. Plant. Res.* **2015**, *2*, 85–89.

55. Kashyap, D.; Siddiqui, Z.A. Effect of silicon dioxide nanoparticles and Rhizobium leguminosarum alone and in combination on the growth and bacterial blight disease complex of pea caused by Meloidogyne incognita and Pseudomonas syringae pv. pisi. *Arch. Phytopathol. Plant Prot.* **2020**, *3*, 1–7. [CrossRef]

56. Adhikari, S.; Kundu, S.; Rao, S.A. Impact of SiO_2 and Mo nanoparticles on seed germination of rice (*Oryza sativa* L.). *Int. J. Agric. Food Sci. Tech.* **2013**, *4*, 809–816.

57. Yinfeng, X.; Bo, L.; Gongsheng, T.A.O.; Qianqian, Z.; Chunxia, Z. Effects of nano-silicon dioxide on photosynthetic fluorescence characteristics of *Indocalamus barbatus* McClure. *J. Nanjing For. Univer.* **2012**, *36*, 59–63.

58. Suriyaprabha, R.; Karunakaran, G.; Yuvakkumar, R.; Rajendran, V.; Kannan, N. Silica nanoparticles for increased silica availability in maize (*Zea mays.* L) seeds under hydroponic conditions. *Curr. Nanosci.* **2012**, *8*, 902–908. [CrossRef]

59. Yuvakkumar, R.; Elango, V.; Rajendran, V.; Kannan, N.S.; Prabu, P. Influence of nanosilica powder on the growth of maize crop (*Zea mays L.*). *Int. J. Green Nanotechnol.* **2011**, *3*, 180–190. [CrossRef]

60. Lu, C.M.; Zhang, C.Y.; Wen, J.Q.; Wu, G.R.; Tao, M.X. Research on the effect of nanometer materials on germination and growth enhancement of *Glycine max* and its mechanism. *Soybean Sci.* **2002**, *21*, 68–172.

61. Zhao, L.; Ortiz, C.; Adeleye, A.S.; Hu, Q.; Zhou, H.; Huang, Y.; Keller, A.A. Metabolomics to detect response of lettuce (Lactuca sativa) to $Cu(OH)_2$ nanopesticides: Oxidative stress response and detoxification mechanisms. *Environ. Sci. Technol.* **2016**, *50*, 9697–9707. [CrossRef] [PubMed]

62. Zhao, L.; Huang, Y.; Adeleye, A.S.; Keller, A.A. Metabolomics reveals $Cu(OH)_2$ nanopesticide-activated anti-oxidative pathways and decreased beneficial antioxidants in spinach leaves. *Environ. Sci. Technol.* **2017**, *51*, 10184–10194. [CrossRef] [PubMed]

63. Yusefi-Tanha, E.; Fallah, S.; Rostamnejadi, A.; Pokhrel, L.R. Zinc oxide nanoparticles (ZnO NPs) as a novel nanofertilizer: Influence on seed yield and antioxidant defense system in soil grown soybean (*Glycine max* cv. Kowsar). *Sci. Total Environ.* **2020**, *738*, 140240. [CrossRef]

64. Singh, N.B.; Amist, N.; Yadav, K.; Singh, D.; Pandey, J.K.; Singh, S.C. Zinc oxide nanoparticles as fertilizer for the germination, growth and metabolism of vegetable crops. *J. Nano Eng. Nano Manuf.* **2013**, *3*, 1–12. [CrossRef]

65. Prasad, T.N.V.K.V.; Sudhakar, P.; Sreenivasulu, Y.; Latha, P.; Munaswamy, V.; Reddy, R.K.; Sreeprasad, T.S.; Sajanlal, P.R.; Pradeep, T. Effect of nanoscale zinc oxide particles on the germination, growth and yield of Peanut. *J. Plant Nutr.* **2012**, *35*, 905–927. [CrossRef]

66. Srivastava, G.; Das, C.K.; Das, A.; Singh, S.K.; Roy, M.; Kim, H.; Sethy, N.; Kumar, A.; Sharma, R.K.; Singh, S.K.; et al. Seed treatment with iron pyrite (FeS_2) nanoparticles increases the production of spinach. *RSC Adv.* **2014**, *4*, 58495–58504. [CrossRef]

67. Rezaei, F.; Moaveni, P.; Mozafari, H. Effect of different concentrations and time of nano TiO_2 spraying on quantitative and qualitative yield of soybean (*Glycine max L.*) at Shahr-e-Qods. *Iran. Biol. Forum.* **2015**, *7*, 957–964.

68. Samadi, N.; Yahyaabadi, S.; Rezayatmand, Z. Effect of TiO_2 and TiO_2 Nanoparticle on Germination, Root and Shoot Length and Photosynthetic Pigments ofMentha Piperita. *Int. J. Plant. Soil Sci.* **2014**, *3*, 408–418. [CrossRef]

69. Reyhaneh, A.; Hassan, F.; Hosseini, M.K. Can Bulk and Nanosized Titanium Dioxide Particles Improve Seed Germination Features of Wheatgrass (*Agropyron desertorum*). *Not. Sci. Biol.* **2013**, *5*, 325–331.

70. Kim, S.W.; Jung, J.H.; Lamsal, K.; Kim, Y.S.; Min, J.S.; Lee, Y.S. Antifungal effects of silver nanoparticles (AgNPs) against various plant pathogenic fungi. *Mycobiology* **2012**, *4*, 53–58. [CrossRef]

71. Gautam, N.; Salaria, N.; Thakur, K.; Kukreja, S.; Yadav, N.; Yadav, R.; Goutam, U. Green silver nanoparticles for phytopathogen control. *Proc. Natl. Acad. Sci. India Sect. B Biol. Sci.* **2020**, *90*, 439–446. [CrossRef]

72. Rodríguez-Serrano, C.; Guzmán-Moreno, J.; Ángeles-Chávez, C.; Rodríguez-González, V.; Ortega-Sigala, J.J.; Ramírez-Santoyo, R.M.; Vidales-Rodríguez, L.E. Biosynthesis of silver nanoparticles by Fusarium scirpi and its potential as antimicrobial agent against uropathogenic Escherichia coli biofilms. *PLoS ONE* **2020**, *15*, e0230275. [CrossRef]

73. Hussein, E.A.M.; Al-Hajry, A.M.; Harraz, F.A.; Ahsan, M.F. Biologically synthesized silver nanoparticles for enhancing tetracycline activity against *Staphylococcus aureus* and *Klebsiella pneumonia*. *Braz. Arch. Biol. Technol.* **2019**, *62*, 25. [CrossRef]

74. Shehzad, A.; Qureshi, M.; Jabeen, S.; Ahmad, R.; Alabdalall, A.H.; Aljafary, M.A.; Al-Suhaimi, E. Synthesis, Characterization and antibacterial activity of silver nanoparticles using Rhazya stricta. *Peer J.* **2018**, *6*, e6086. [CrossRef]

75. Mohanta, Y.K.; Panda, S.K.; Bastia, A.K.; Mohanta, T.K. Biosynthesis of silver nanoparticles from protium serratum and Potential Impacts on food safety and control. *Front. Microbiol.* **2017**. [CrossRef] [PubMed]

76. Abdelmalek, G.A.M.; Salaheldin, T.A. Silver nanoparticles as a potent fungicide for citrus phytopathogenic fungi. *J. Nanomed. Res.* **2016**, *3*, 1–8.

77. Krishnaraj, C.; Ramachandran, R.; Mohan, K.; Kalaichelvan, P. Optimization for rapid synthesis of silver nanoparticles and its effect on phytopathogenic fungi. *Spectrochim. Acta A* **2012**, *93*, 95–99. [CrossRef] [PubMed]

78. Jo, Y.-K.; Kim, B.H.; Jung, G. Antifungal activity of silver ions and nanoparticles on phytopathogenic fungi. *Plant. Dis.* **2009**, *93*, 1037–1043. [CrossRef] [PubMed]

79. Shahryari, F.; Rabiei, Z.; Sadighian, S. Antibacterial activity of synthesized silver nanoparticles by sumac aqueous extract and silver-chitosan nanocomposite against *Pseudomonas syringae* pv. *syringae*. *J. Plant. Pathol.* **2020**, *102*, 469–475. [CrossRef]

80. Divya, K.; Vijayan, S.; George, T.K.; Jisha, M. Antimicrobial properties of chitosan nanoparticles: Mode of action and factors affecting activity. *Fibers Polym.* **2017**, *18*, 221–230. [CrossRef]

81. Xing, K.; Shen, X.; Zhu, X.; Ju, X.; Miao, X.; Tian, J.; Feng, Z.; Peng, X.; Jiang, J.; Qin, S. Synthesis and in vitro antifungal efcacy of oleoyl-chitosan nanoparticles against plant pathogenic fungi. *Int. J. Biol. Macromol.* **2016**, *82*, 830–836. [CrossRef]

82. Dang, H.; Fawcett, D.G.; Poinern, J. Green synthesis of gold nanoparticles from waste macadamia nuts shells and their antimicrobial activity against Escherichia coli and *Staphylococcus epidermis*. *Int. J. Res. Med. Sci.* **2019**. [CrossRef]

83. Attar, A.; Yapaoz, M.A. Biomimetic synthesis, characterization and antibacterial efficacy of ZnO and Au nanoparticles using Echinacea flower extract precursor. *Mater. Res. Express.* **2018**, *5*, 5. [CrossRef]

84. Jayaseelan, C.; Ramkumar, R.; Rahuman, A.A.; Perumal, P. Green synthesis of gold nanoparticles using seed aqueous extract of Abelmoschus esculentus and its antifungal activity. *Ind. Crops Prod.* **2013**, *45*, 423–429. [CrossRef]

85. Fan, Q.; Liao, Y.Y.; Kunwar, S.; Da Silva, S.; Young, M.; Santra, S.; Paret, M.L. Antibacterial effect of copper composites against Xanthomonas euvesicatoria. *Crop Prot.* **2020**, *139*, 105366. [CrossRef]

86. Huang, S.; Wang, L.; Liu, L.; Hou, Y.; Li, L. Nanotechnology in agriculture, livestock, and aquaculture in China. A review. *Agron. Sustain. Dev.* **2015**, *35*, 369–400. [CrossRef]

87. Giannousi, K.; Avramidis, I.; Dendrinou-Samara, C. Synthesis, characterization and evaluation of copper based nanoparticles as agrochemicals against *Phytophthora infestans*. *RSC Adv.* **2013**, *3*, 21743–21752. [CrossRef]

88. Sharma, N.; Jandaik, S.; Kumar, S.; Chitkara, M.; Sandhu, I.S. Synthesis, characterisation and antimicrobial activity of manganese- and iron-doped zinc oxide nanoparticles. *J. Exp. Nanosci.* **2015**, *11*, 54–71. [CrossRef]

89. Derbalah, A.; Shenashen, M.; Hamza, A.; Mohamed, A.; El Safty, S. Antifungal activity of fabricated mesoporous silica nanoparticles against early blight of tomato. *Egypt. J. Basic Appl. Sci.* **2018**, *5*, 145–150. [CrossRef]

90. Akpinar, I.; Sar, T.; Unal, M. Antifungal effects of silicon dioxide nanoparticles (SiO$_2$ NPs) against various plant pathogenic fungi. In *International Workshop: Plant Health: Challenges and Solutions*; Tar, M., Baştaş, K.K., Eds.; Frontiers Media SA: Antalya, Turkey, 2017; ISBN 978-2-88945-218-7.

91. Park, H.J.; Kim, S.H.; Kim, H.J.; Choi, S.H. A new composition of nano sized silica-silver for control of various plant diseases. *Plant Pathol.* **2006**, *22*, 295–302. [CrossRef]

92. Jamdagni, P.; Khatri, P.; Rana, J.S. Green synthesis of zinc oxide nanoparticles using flower extract of *Nyctanthesarbor-tristis* and their antifungal activity. *J. King Saud Uni. Sci.* **2016**. [CrossRef]

93. Navale, G.R.; Thripuranthaka, M.; Late, D.J.; Shinde, S.S. Antimicrobial Activity of ZnO Nanoparticles against Pathogenic Bacteria and Fungi. *JSM Nanotechnol. Nanomed.* **2015**, *3*, 10–33.

94. Rajiv, P.; Rajeshwari, S.; Venckatesh, R. Bio-Fabrication of zinc oxide nanoparticles using leaf extract of *Parthenium hysterophorus* L. and its size-dependent antifungal activity against plant fungal pathogens. *Spectrochim. Acta A* **2013**, *112*, 384–387. [CrossRef] [PubMed]

95. Gunalan, S.; Sivaraj, R.; Rajendran, V. Green synthesized ZnO nanoparticles against bacterial and fungal pathogens. *Prog. Nat. Sci. Mater.* **2012**, *22*, 693–700. [CrossRef]

96. Dimkpa, C.O.; McLean, J.E.; Britt, D.W.; Anderson, A.J. Antifungal activity of ZnO nanoparticles and their interactive effect with a biocontrol bacterium on growth antagonism of the plant pathogen Fusarium graminearum. *Biometals* **2013**, *26*, 913–924. [CrossRef] [PubMed]

97. Jayaseelan, C.; Rahuman, A.A.; Kirthi, A.V.; Marimuthu, S.; Santhoshkumar, T.; Bagavan, A.; Rao, K.V.B. Novel microbial route to synthesize ZnO nanoparticles using Aeromonas hydrophila and their activity against pathogenic bacteria and fungi. *Spectrochim. Acta Part A Mol. Biomol. Spectrosc.* **2012**, *90*, 78–84. [CrossRef]

98. Sar, T.; Akpinar, I.; Unal, M. Inhibitory effect of antifungal activity of Titanium Dioxide (TiO$_2$) nanoparticles on some pathogenic *Fusarium* isolates. In Proceedings of the International Workshop Plant Health: Challenges and Solutions, Antalya, Turkey, 23–28 April 2017. 78p.

99. Hamza, A.; El-Mogazy, S.; Derbalah, A. Fenton reagent and titanium dioxide nanoparticles as antifungal agents to control leaf spot of sugar beet under field conditions. *J. Plant Prot. Res.* **2016**, *56*, 270–278. [CrossRef]

100. Ardakani, A.S. Toxicity of silver, titanium and silicon nanoparticles on the root-knot nematode, Meloidogyne incognita, and growth parameters of tomato. *Nematology* **2013**, 1–7. [CrossRef]

101. Kasemets, K.; Ivask, A.; Dubourgier, H.C.; Kahru, A. Toxicity of nanoparticles of ZnO, CuO and TiO$_2$ to yeast *Saccharomyces cerevisiae*. *Toxicol. Vitr.* **2009**, *23*, 1116–1122. [CrossRef] [PubMed]

102. Cui, H.; Zhang, P.; Gu, W.; Jiang, J. Application of anatasa TiO$_2$ sol derived from peroxotitannic acid in crop diseases control and growth regulation. *NSTI Nanotech.* **2009**, *2*, 286–289.

103. Slavin, Y.N.; Asnis, J.; Häfeli, U.O.; Bach, H. Metal nanoparticles: Understanding the mechanisms behind antibacterial activity. *J. NanobioTechnol.* **2017**, *15*, 65. [CrossRef]

104. Singh, J.; Vishwakarma, K.; Ramawat, N.; Rai, P.; Singh, V.K.; Mishra, R.K.; Kumar, V.; Tripathi, D.K.; Sharma, S. Nanomaterials and microbes' interactions: A contemporary overview. *3 Biotech* **2019**, *9*, 1–4. [CrossRef]

105. Kasprowicz, M.J.; Kozioł, M.; Gorczyca, A. The effect of silver nanoparticles on phytopathogenic spores of Fusarium culmorum. *Can. J. Microbiol.* **2010**, *56*, 247–253. [CrossRef] [PubMed]

106. De Paz, L.E.; Resin, A.; Howard, K.A.; Sutherland, D.S.; Wejse, P.L. Antimicrobial effect of chitosan nanoparticles on Streptococcus mutans biofilms. *Appl. Environ. Microbiol.* **2011**, *77*, 3892–3895. [CrossRef]

107. Aljabali, A.A.; Akkam, Y.; Al Zoubi, M.S.; Al-Batayneh, K.M.; Al-Trad, B.; Abo Alrob, O.; Alkilany, A.M.; Benamara, M.; Evans, D.J. Synthesis of gold nanoparticles using leaf extract of Ziziphus zizyphus and their antimicrobial activity. *Nanomaterials* **2018**, *3*, 174. [CrossRef]

108. Yuan, C.G.; Huo, C.; Gui, B.; Cao, W.P. Green synthesis of gold nanoparticles using Citrus maxima peel extract and their catalytic/antibacterial activities. *IET NanobioTechnol.* **2016**, *11*, 523–530. [CrossRef]

109. Khan, R.A.; Tang, Y.; Naz, I.; Alam, S.S.; Wang, W.; Ahmad, M.; Najeeb, S.; Ali, A.; Rao, C.; Li, Y.; et al. Management of Ralstonia solanacearum in tomato using ZnO nanoparticles synthesized through Matricaria chamomilla. *Plant Disease.* **2021**, *28*. [CrossRef]

110. Kairyte, K.; Kadys, A.; Luksiene, Z. Antibacterial and antifungal activity of photoactivated ZnO nanoparticles in suspension. *J. Photochem. Photobiol.* **2013**, *128*, 78–84. [CrossRef]

111. Patra, P.; Mitra, S.; Debnath, N.; Goswami, N. Biochemical-Biophysical-and microarray-based antifungal evaluation of the buffer-mediated synthesized nano zinc oxide: An in vivo and in vitro toxicity study. *Langmuir* **2012**, *28*, 16966–16978. [CrossRef] [PubMed]

112. Wani, A.H.; Shah, M.A. unique and profound effect of MgO and ZnO nanoparticles on some plant pathogenic fungi. *J. Appl. Pharm. Sci.* **2012**, *2*, 40–44.

113. He, L.; Liu, Y.; Mustapha, A.; Lin, M. Antifungal activity of zinc oxide nanoparticles against *Botrytis cinerea* and *Penicillium expansum*. *Microbiol. Res.* **2011**, *166*, 207–215. [CrossRef] [PubMed]

114. Sharma, D.; Rajput, J.; Kaith, B.S.; Kaur, M.; Sharma, S. Synthesis of ZnO nanoparticles and study of their antibacterial and antifungal properties. *Thin Solid Films* **2010**, *519*, 1224–1229. [CrossRef]

115. Norman, D.J.; Chen, J. Effect of foliar application of titanium dioxide on bacterial blight of geranium and Xanthomonas leaf spot of poinsettia. *Hortic. Sci.* **2011**, *46*, 426–428. [CrossRef]

116. Sanzari, I.; Leone, A.; Ambrosone, A. Nanotechnology in Plant Science: To Make a Long Story Short. *Front. Bioeng. Biotechnol.* **2019**, *7*, 120. [CrossRef]

117. Naushad, M. Surfactant assisted nano-composite cation exchanger: Development, characterization and applications for the removal of toxic Pb^{2+} from aqueous medium. *Chem. Eng. J.* **2014**, *235*, 100–108. [CrossRef]

118. Albadarin, A.B.; Collins, M.N.; Naushad, M.; Shirazian, S.; Walker, G.; Mangwandi, C. Activated lignin-chitosan extruded blends for efficient adsorption of methylene blue. *Chem. Eng. J.* **2017**, *307*, 264–272. [CrossRef]

119. Kushwah, K.S.; Patel, S. Effect of Titanium Dioxide Nanoparticles (TiO_2 NPs) on Faba bean (*Vicia faba* L.) and Induced Asynaptic Mutation: A Meiotic Study. *J. Plant. Growth Regul.* **2020**, *39*, 1107–1118. [CrossRef]

120. Gohari, G.; Mohammadi, A.; Akbari, A.; Panahirad, S.; Dadpour, M.R.; Fotopoulos, V.; Kimura, S. Titanium dioxide nanoparticles (TiO_2 NPs) promote growth and ameliorate salinity stress effects on essential oil profile and biochemical attributes of *Dracocephalum moldavica*. *Sci. Rep.* **2020**, *10*, 912. [CrossRef]

121. Patlolla, A.K.; Berry, A.; May, L.; Tchounwou, P.B. Genotoxicity of Ag NPs in Vicia faba: A pilot study on the environmental monitoring of NPs. *Int. J.Environ.Res. Public Health* **2012**, *9*, 1649–1662. [CrossRef] [PubMed]

122. Lopez-Moreno, M.L.; de la Rosa, G.; Hernández-Viezcas, J.; Castillo-Michel, H.; Botez, C.E.; Peralta-Videa, J.R.; Gardea-Torresdey, J.L. Evidence of the Differential Biotransformation and Geno-toxicity of ZnO and CeO_2 NPs on Soybean (*Glycine max*) Plants. *Environ. Sci. Technol.* **2010**, *44*, 7315–7320. [CrossRef] [PubMed]

123. Manjunatha, S.B.; Biradar, D.P.; Aladakatti, Y.R. Nanotechnology and its applications in agriculture: A review. *J. Farm Sci.* **2016**, *29*, 1–3.

applied sciences

Article

Estimation of Risk to the Eco-Environment and Human Health of Using Heavy Metals in the Uttarakhand Himalaya, India

Amit Kumar [1], Marina Cabral-Pinto [2,*], Amit Kumar [3,*], Munesh Kumar [4] and Pedro A. Dinis [5]

1   School of Hydrology and Water Resources, Nanjing University of Information Science and Technology, Nanjing 210044, China; amitkdah@nuist.edu.cn or amit.agl09@gmail.com
2   Geobiotec Research Centre, Department of Geosciences, University of Aveiro, 3810-193 Aveiro, Portugal
3   Central MugaEri Research and Training Institute, Central Silk Board, Jorhat, Assam 785000, India
4   Department of Forestry and Natural Resources, H.N.B. Garhwal University (A Central University), Srinagar Garhwal 249161, Uttarakhand, India; munesh.hnbgu@gmail.com
5   Marine and Environmental Sciences Centre (MARE), Department of Earth Sciences, University of Coimbra, Rua Sílvio Lima, 3030-790 Coimbra, Portugal; pdinis@dct.uc.pt
*   Correspondence: marinacp@ua.pt (M.C.-P.); amitkumar.csb@gov.in (A.K.); Tel.: +351-964332189 (M.C.-P.); +91-8130952665 (A.K.)

Received: 31 August 2020; Accepted: 1 October 2020; Published: 12 October 2020

Abstract: In the modern era, due to the rapid increase in urbanization and industrialization in the vicinity of the Himalayas, heavy metals contamination in soil has become a key priority for researchers working globally; however, evaluation of the human and ecological risks mainly in hilly areas remains limited. In this study, we analyzed indices like the contamination factor (CF), degree of contamination (DC), enrichment factor (EF), geochemical index (I_{geo}), pollution ecological risk index (PERI), and pollution load index (PLI), along with cancer risk (CR) and hazard indices (HI), to ascertain the eco-environmental and human risks of using heavy metals in datasets collected from 168 sampling locations in Uttarakhand, India. The evaluation calculated of I_{geo}, EF, and CF suggests that represented soil samples were moderately contaminated and highly augmented with Rb, while PERI (75.56) advocates a low ecological risk. Further, PLI and DC (PLI: 1.26; DC: 36.66) show a possible health risk for the native population in the vicinity of the studied catchment. The hazard index (HI) is estimated greater than 1 (HI > 1) for Cr and Mn, representing a possible risk for cancer. However, adults are free from cancer risk, and other studied elements have been reported as noncarcinogenic. This assessment gives important information to policymakers, environmentalists, and foresters for taking mitigation measures in advance to mitigate the potential future risk of soil pollution on humans, ecology, and the environment.

Keywords: risk; heavy metal; human health; pollution indices; soil

---

1. Introduction

Due to the ubiquitous bioaccumulation of toxic heavy metals, even at the trace level, and the persistent concentration increase in soils and possible uptake through groundwater, atmosphere, crops, and the food chain, these elements are hazardous for human health [1–6]. Thus, heavy metals contamination in soil has been drawing much attention globally [7–10]. Generally, heavy metals in natural resources can be either natural (lithogenic inputs via weathering rocks) or anthropogenic in origin [11–14]. The change in climate conditions and socio-development features may amplify the inclusion of heavy metals in soils either by weathering or contaminant inputs such as surface run-off, wastewater, sewage, and effluent, and has become a common but intense issue in various ecosystems [13,15–18]. These environmental problems are widespread in India [19–22]. Generally, long-term exposure to heavy metals profoundly creates health risks. Nonetheless, heavy metals do not show any adverse effect on human health and the environment if they are present at safe levels. If the concentrations of heavy metals exceed the safe limits, then they often cause acute health risks to mankind with severe consequences [9,10,16]. There are some trace elements that are essential for humans, such as iron, manganese, zinc, aluminum, lead, and cadmium; however, they may exert toxicity under pathological or artificially harsh conditions of exposure to excess levels [9–12,23,24], and references therein [25–29].

Soil characteristics are of considerable importance in holding the fertility of soil and balancing the nutrients requirements of different food chains [30]. The key information and mechanistic understanding of the outcome of various trace metals [31] and their dynamics are still under investigation in various ecosystems [32]. Furthermore, agriculture and the urban management of soil force change in the soil structure, function, and pedological properties [3,33,34]. In view of this, screening tools such as spectroscopic techniques (X-ray fluorescence (XRF)) and infrared diffuse reflectance techniques to understand the possible change in soil properties have been developed in recent years [35,36] and are used to investigate the elemental composition of the soil [37,38]. In this direction, important indices such as the cancer risk (CR), enrichment factor (EF), contamination factor (CF), geochemical index (I_{geo}), and hazard indices (HI) have also been used for quantifying the soil contamination and risk to the eco-environment and human health at regional and/or global levels [9,10,34,39–41]. Assessment of these important indices needs the existing background concentration (BC value) of individual heavy metals in the earth's crust. Thus, investigation for the assessment of the background concentration of these heavy metals in local/regional, national, continental, and global soils under different land-use systems was established [42–49]. However, the elemental concentration may vary in different land-use/regions due to variations in various controlling factors that contribute to soil formation and spatial distribution [43]. Therefore, it is of utmost importance to assess the local background concentration (BC) value of targeted heavy metal/metals in soil of specific region/land-use. This parameter is crucial to estimate the magnitude of contamination and its ameliorative measures for sustainable growth and development of the natural/anthropogenic ecosystem [11,12,44–52]. In India, the same trends have also been attempted by various researchers in different regions [19–22,53–59], but these studies are limited in the hills of the Indian Himalayas. Therefore, the present investigation was conducted to fill a knowledge gap that undermines the assessment of potential ecological and human health (carcinogenic) risks in the high hills of Uttarakhand, India. The results would be helpful for policy-makers and environmentalists to reduce the risk by making strategic mitigation plans in the future.

2. Materials and Methods

2.1. Study Site

Uttarakhand (28 44′, 31 28′ N to 77 35′, 81 01′ E) is administratively divided into 13 districts, having 71% (37,999.53 km^2) of forest cover compared to the total geographical area (53,483 km^2 or 20,650 square miles), and shares its border as the international boundary with the northern region of China and eastern region of Nepal (Figure 1). The majority of the Uttarakhand population (10.11 million) lives in rural areas. Forest is the major asset of the state. The average annual precipitation is ~1550 mm. The climatic conditions are majorly temperate; however, tropical and subtropical climatic conditions exist only in plains and some of the foothills of the state. Chauhan et al. [60] classify the state into 2 high-altitude regions, i.e., lower (2400–4500 m above mean sea level and upper (>4500 m); 2 hilly regions, i.e., lower (300–600 m) and upper (600–2400 m); and a single terrain region (<300 m). The Uttarakhand climate is a cold, humid, and temperate type having altitude variation [60]. The soil fertility status in the state varies from low to medium among brown hill, mountain meadow, red loamy, and sub-mountain soils. Most of the soils in the high-altitude region of Uttarakhand are acidic (pH < 7) in nature. Agricultural crops (e.g., rice, wheat, and sugarcane) and horticulture crops (tomato, cauliflower, cabbage, etc.) are the main crops in the high hills of the Himalayas. In the most populated altitudinal range (i.e., 1000–2000 m above MSL), there are oak (*Quercus spp.*), pine (*Pinus roxburghii*), banj (*Quercus leucotrichophora*), and buras (*Rhododendron arboreum*). Coniferous forests, e.g., deodar (*Cedrus deodara*) and Fir (*Abies species*), form the dominant forest vegetation at high altitudinal ranges (>2000 m) of the state [61–63].

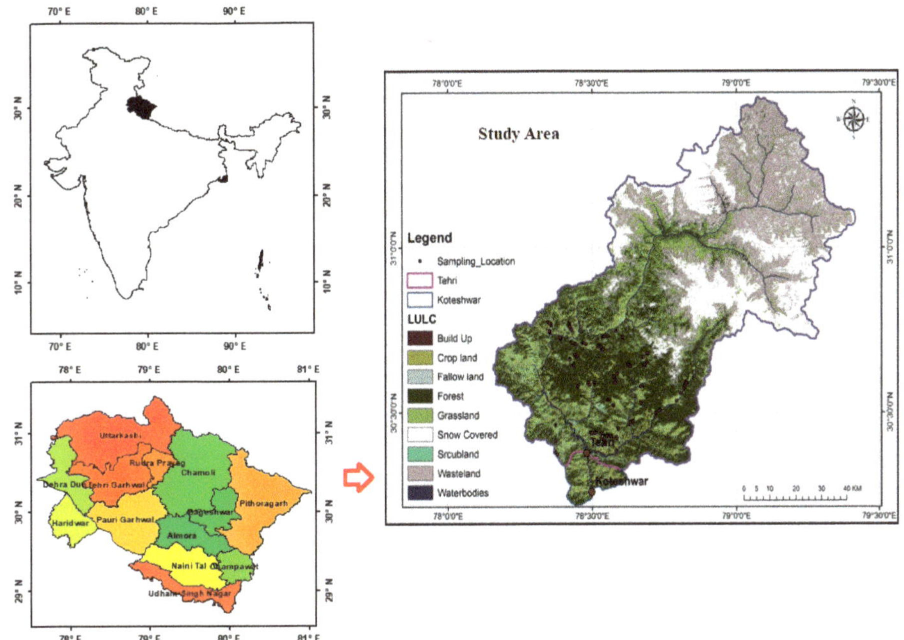

Figure 1. Study area in the high hills of Uttarakhand Himalaya, India.

2.2. Geology and Soils

From north to south (Figure 2a), the Himalaya in Uttarakhand includes four tectono-stratigraphic units [64,65]. (1) The Tethys, or Tibetan, Himalayan Zone comprising Palaeozoic to Eocene units deposited on the northern margin of the Indian plate [66,67]; (2) the Higher, or Greater, Himalaya Zone comprising Proterozoic crystalline rocks intruded by leucogranites [68,69]; (3) the Lesser Himalaya Zone with non-metamorphosed or weakly metamorphosed Indian continental crust and its sedimentary cover [69], intruded by Proterozoic plutons [70], along with Paleogene foreland deposits [71]; (4) the Outer Himalaya Zone comprising a thick Palaeocene to Quaternary sedimentary succession composed mainly of continental units derived from the Himalayan orogen [72–75].

The studied soils cover geological units of the Lesser Himalaya and Greater Himalaya zones (Figure 2A,B). In the investigated region, the Greater Himalaya is mainly composed of micaceous schists, gneisses, calc-silicate gneiss, and locally metabasic rocks [69]. The Lesser Himalaya units are from its inner metasedimentary belt (diverse siliciclastic and carbonate metasedimentary and volcanic units) and outer sedimentary belt (diamictite, sandstone, slate, carbonates, chert/quartzite) [69,74].

Figure 2B presents the adapted soil cartography of Uttarakhand, according to the FAO/UNESCO [75]. The soil map reveals the strong link between geology, relief, and soil-type (Figure 2B). The higher-elevation areas with crystalline rocks of the Greater Himalaya are occupied by lithosols, the lower Lesser Himalaya dominated by sedimentary units displays cambisols, and the transitional zone between these two geological units can be covered by either cambisols or lithosols.

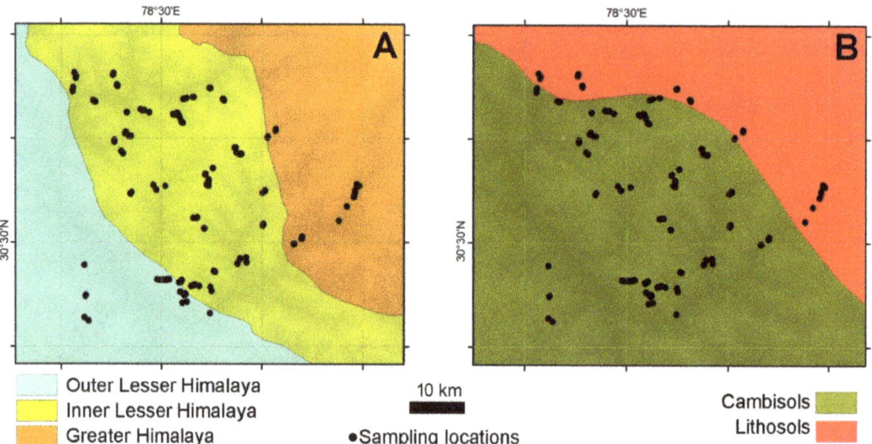

Figure 2. (**A**) Simplified geological (based on Mukherjee [76–79]) and (**B**) soil sketch of the studied area, based on FAO [75].

2.3. Soil Sampling and Analysis

Sampling of the soil was conducted in the Tehri Garhwal and Urrarkashi forest division of the Uttarakhand state of India in two stages (I-Tehri Garhwal; II-Urrarkashi forest division). A total of 168 randomized soil samples at a depth of 30 cm (1 in. = 2.54 cm) were collected following recommendations by the forest research institute (FRI, Dehradun, India) from different altitudinal ranges (Table 1) under the various forest species, as shown in Figure 1. Snow (18%) and wasteland (27%) of the catchment area were not considered for sampling, due to infeasibility conditions [62]. The coordinates and altitude of every individual sampling point were recorded by GPS (Garmin 76CSx, Garmin International Inc. Kansas, U.S.A). While collecting soil samples, a 0.50×0.50 m^2 area of 30 cm depth was dug out using a soil auger at every location to collect soil (~500 g). Further, collected soil was mixed thoroughly; root debris, gravel, etc. were opted out from the samples; and then a composite sample was stored in polythene bags for further analysis in the laboratory with proper labeling to avoid errors. Individual soil samples along with the labeling were air-dried and then finely ground and sieved using a stainless-steel sieve of 100 mesh size. Further, elemental analyses, viz. Sr, Zn, Rb, Cu, Ni, Ca, Fe, Mn, Si, Al, Mg Cr, K, S, P, and Na, were analyzed using X-ray fluorescence (XRF, model: 54 pioneer, Bruker, 2008) [35]. The parametric statistical analysis (mean ± SD) of individual heavy metals results was calculated and is listed in Table 1.

2.4. Eco-Environmental Risk Assessment

Eco-environmental risks were assessed using geostatistical indices to determine the degree and cause (e.g., anthropogenic or geogenic) of heavy metals contamination (Table 2). Indices like the degree of contamination (DC), contamination factor (CF), enrichment factor (EF), pollution load index (PLI), geo-accumulation index (I_{geo}), pollution ecological risk index (PERI), and potential contamination index (PCI) were estimated using recommended methods [34,39,63]. In India, region-specific background heavy metals composition values are not available for the studied regions; therefore, relevant values provided by Taylor and McLennan [61] were used to evaluate the indices followed by interpretation.

Table 1. Chemical elements concentration of soil (mean ± SD) vs. altitudes.

Element (mg/kg)	Concentration in Upper Continental Crust (B) (mg/kg)	<1000 m	1001–1500 m	1501–2000 m	2001–2500 m	>2500 m
Sodium (Na)	28,900	3400 ± 297	1746 ± 695	2578 ± 1443	2614 ± 601	1767 ± 736
Magnesium (Mg)	13,300	9780 ± 1018	5600 ± 1344	8141 ± 3768	8036 ± 2770	8230 ± 2402
Aluminum (Al)	80,400	59,850 ± 4596	62,181 ± 8452	68,326 ± 8116	65,887 ± 7617	61,367 ± 3099
Silicon (Si)	308,000	228,000 ± 2828	214,667 ± 27,390	205,894 ± 27,718	196,875 ± 28,175	213,666 ± 20,502
Phosphorus (P)	700	882 ± 157	790 ± 345	974 ± 515	1442 ± 885	8912 ± 89
Sulfur (S)	500	293 ± 57	337 ± 140	426 ± 186	667 ± 313	406 ± 55
Potassium (K)	28,000	25,350 ± 5303	25,819 ± 6808	26,974 ± 6739	23,525 ± 4364	19,467 ± 4196
Calcium (Ca)	30,000	7765 ± 8676	10,860 ± 16,511	8621 ± 8314	11,051 ± 6576	5703 ± 3854
Chromium (Cr)	35	173 ± 16	180 ± 34	171 ± 53	198 ± 34	201 ± 16
Manganese (Mn)	600	1342 ± 1100	923 ± 318	965 ± 663	2069 ± 2211	1087 ± 643
Iron (Fe)	35,000	67,350 ± 6717	54,233 ± 16,734	62,716 ± 27,569	67,181 ± 22,068	55,633 ± 13,250
Nickel (Ni)	20	99 ± 22	106 ± 37	104 ± 31	128 ± 23	112 ± 6
Copper (Cu)	25	179 ± 55	155 ± 39	162 ± 73	152 ± 31	133 ± 18
Zinc (Zn)	71	143 ± 15	156 ± 46	153 ± 55	197 ± 62	172 ± 28
Rubidium (Rb)	112	564 ± 300	729 ± 390	956 ± 1038	592 ± 266	464 ± 246
Strontium (Sr)	350	319 ± 39	215 ± 89	235 ± 79	353 ± 85	250 ± 107
No, of samples collected		6	58	51	45	8

Table 2. Indices used to assess ecological risk for study area.

Indices	Information	Equations	Pollution Classification	Ref.
I_{geo}	Determines the extent of soil contamination from a comparison of the concentration of elements in the sil sample and the upper continental crust (UCC)	$I_{geo} = log_2 \left[\frac{C_n}{1.5 \times B_n} \right]$	I_{geo} can be classified as: $I_{geo} \leq 0$: Uncontaminated; $0 < I_{geo} \leq 1$: Uncontaminated to moderately contaminated; $1 < I_{geo} \leq 2$: Moderately contaminated; $2 < I_{geo} \leq 3$: Moderately to highly contaminated; $3 < I_{geo} \leq 4$: Highly to extremely contaminated; $I_{geo} > 5$: Extremely contaminated	[65,66]
EF	Standardizes the concentration of analyzed element against a reference element in the soil and the UCC	$EF = \left[\frac{C_n^{sample} / Al_{sample}}{C_n^{crust} / Al_{crust}} \right]$	EF classified as: $0 < EF < 1$: No enrichment; $1 < EF < 3$: Slight enrichment; $3 < EF < 5$: Reasonable enrichment; $5 < EF < 10$: Reasonably high enrichment; $10 < EF < 25$: High enrichment; $25 < FE < 50$: Very high enrichment; and $EF > 50$: Extremely high enrichment.	[63,80]
CF	Represents the effect or contribution of an individual element in soil contamination	$CF = \frac{C_n}{B_n}$	CF classified as: $CF < 1$: Low contamination; $1 < CF < 3$: Reasonable contamination; $3 < CF > 6$: Considerable contamination; and $CF > 6$: High or very high contamination	[60,80]
PLI	The product of CF in the soil sample	$PLI = (CF_1 \times CF_2 \times CF_3 \ldots \ldots CF_n)^{1/n}$	PLI classified as: $PLI > 1$ indicates the presence of pollution, whereas PLI < 1 indicates no elemental pollution	[80,81]
DC	The sum of CF of eight elements (Cr, Mn, Fe, Ni, Cu, Zn, Rb, Sr) considered for the study (i.e., cumulative effect of the heavy metals in soil contamination)	$DC = \sum_{i=1}^{i=n} CF$	DC classified as: $DC < 8$: Low contamination; $8 < DC < 16$: Reasonable contamination; $16 < DC < 32$: Considerable contamination; and $DC > 32$: High degree of contamination (i.e., serious anthropogenic impacts)	[82]
PCI	Used in forest ecosystems where heavy metals exist as complex mixtures with spatiotemporal variability.	$PCI = \frac{C_n^{max}}{B_n}$	PCI classified as: $PCI < 1$: Low contamination; $1 < PCI < 3$: Reasonable contamination; and $PCI > 3$: Severe or very severe contamination	[80,83]
PERI	Provides an indication of major contamination agents and the identification of sites where studies could prioritize. Used to assess the degree of elemental contamination in response to their toxic effect or risk to the environment	$PERI = \sum P_n$ $P_n = T_n \times CF_n$	Pn classified as: $Pn < 40$: Low potential ecological risk; $40 < Pn < 80$: Reasonable ecological risk; $80 < Pn < 160$: Considerable ecological risk; $160 < Pn < 320$: High ecological risk; and $Pn > 320$: Very high ecological risk. Moreover, PERI classified as: $PERI < 95$: Low potential ecological risk; $95 < PERI < 190$: Moderate ecological risk; $190 < PERI < 380$: Considerable ecological risk; and $PERI > 380$: Very high ecological risk	[80,82]

where C_n: Concentration of an analyzed nth element in the sample (C^{sample}) and upper continental crust (C^{crust}); B_n: BC value of nth element of continental crust; n: Number of elements considered for PLI evaluation; C_n(max): Maximum concentration of nth element found in all collected soil samples; Pn: Potential ecological risk factor of individual heavy metal, which can be evaluated using Tn (toxic response factor) and CF. The standardized value of Tn for metals like Cr, Mn, Ni, Cu, and Zn are 2, 1, 5, 5, and 1, respectively. Tn has been used to identify the toxic response of individual metal, while PERI gives the cumulative effect of five metals (Cr, Mn, Ni, Cu, and Zn) in the study catchment. For the calculation of EF, Al was chosen for normalization to a conservative geogenic element.

2.5. Human Health Risk Assessment

The human health risk assessment was estimated with consideration of direct exposure of toxic heavy metals to adults and children through soils. The carcinogenic and noncarcinogenic risks were estimated as per the methodology of USEPA ([84]; Table 3). Chronic daily intake (CDI) was established for heavy metals exposure through ingestion (CDI_{ingest}), inhalation (CDI_{inha}) and dermal contact (CDI_{dermal}) by adopting the equations given below [84,85]:

$$CDI_{ingest} = \frac{C \times IR \times EF \times ED \times CF}{BW \times AT} \tag{1}$$

$$CDI_{inhalation} = \frac{C \times InhR \times ET \times EF \times ED}{PEF \times BW \times AT} \tag{2}$$

$$CDI_{dermal} = \frac{C \times SA \times AF \times ABS \times EF \times ED \times CF}{BW \times AT} \tag{3}$$

where C denotes the concentration of heavy metals available in soil (mg kg^{-1}). Noncancer risk is represented in terms of hazard index (HI) for multiple substances and/or exposure pathways [34]. HI is the sum of the hazard quotient (HQ), for each element and each pathway, and if HI < 1, there is a very low chance of noncarcinogenic risk. The other terms are explained in Table 3.

Table 3. Reference values adopted for the determination of chronic daily intake (CDI) based on USEPA [84].

Parameter	Units	Adult	Children
IR	mg day^{-1}	100	200
EF	Days year^{-1}	312	312
ED	Years	35	6
BW	kg	70	15
ATnc	Days	365×35	365×6
ATc	Days	365×70	365×70
CF	mg day^{-1}	10^{-6}	10^{-6}
SA	cm^2	6032	2373
AF	mg cm^{-2}	0.07	0.2
ABS	Unitless	0.001	0.001
InhR	m^3 h^{-1}	1.56	1.2
ET	h day^{-1}	8	4
PEF	m^3 kg^{-1}	1.36×10^9	1.36×10^9

where IR: Ingestion rate of soil; BW: Body weight; EF: Exposure frequency; ED: Exposure duration; ATc: Av. time for carcinogenic risk; ATnc: Av. time for noncarcinogenic risk; SA: Skin surface area available for contact; CF: Conversion factor; AF: Soil-to-skin adherence factor; ABS: Absorption factor; InhR: Inhalation rate; ET: Exposure time; PEF: Particle emission factor.

3. Results and Discussion

Heavy metals enter into the terrestrial environment through household wastes, agricultural and anthropogenic processes, the weathering of rocks, etc. where accumulation occurs in the air, soil, and water, while its excess concentration to ecological and human health is a more serious problem in the modern era than water and air pollution [19,34]. These problems are because heavy metals and/or metalloids are generally firmly bound by the organic components available in the topmost soil profile [20]. In general, heavy metals available in low or trace amounts are important for flora and fauna diversity, while high concentrations suppress its growth and finally cause plant decay.

Table 4. Background concentration of individual elements and calculation of eco-environmental indices.

Elements		PCI		CF		EF (Mean ± SD)		I_{geo} (Mean ± SD)	
Na	0.24	low contamination	**0.08**	low contamination	0.10 ± 0.05	no enrichment	−4.38 ± 0.63	uncontaminated
Mg	1.40	reasonable contamination	0.55	low contamination	0.68 ± 0.27	no enrichment	−1.55 ± 0.54	uncontaminated
Al	1.01	reasonable contamination	0.80	low contamination	1.00 ± 0.05	slight enrichment	−0.90 ± 0.18	uncontaminated
Si	0.84	low contamination	0.68	low contamination	0.85 ± 0.18	no enrichment	−1.17 ± 0.20	uncontaminated
P	4.11	severe or very severe contamination	1.41	reasonable contamination	1.84 ± 1.11	slight enrichment	−0.24 ± 0.76	uncontaminated
S	2.46	reasonable contamination	0.89	low contamination	1.15 ± 0.67	slight enrichment	−0.91 ± 0.73	uncontaminated
K	1.36	reasonable contamination	0.89	low contamination	1.12 ± 0.23	slight enrichment	−0.78 ± 0.37	uncontaminated
Ca	2.34	reasonable contamination	0.34	low contamination	0.42 ± 0.51	no enrichment	−2.71 ± 1.14	uncontaminated
Cr	8.34	severe contamination	5.34	considerable contamination	6.55 ± 1.49	reasonably high enrichment	0.39 ± 0.41	tends to moderately contaminated
Mn	14.83	very severe contamination	2.17	reasonable contamination	2.66 ± 2.85	slight enrichment	1.76 ± 0.34	moderately contaminated
Fe	4.09	severe contamination	1.75	reasonable contamination	2.15 ± 0.73	slight enrichment	0.12 ± 0.92	tends to moderately contaminated
Ni	8.55	severe contamination	5.68	considerable contamination	6.95 ± 1.93	reasonably high enrichment	0.11 ± 0.61	tends to moderately contaminated
Cu	16.60	very severe contamination	6.39	high or very high contamination	7.82 ± 2.50	reasonably high enrichment	1.86 ± 0.38	moderately contaminated
Zn	5.59	severe contamination	2.37	reasonable contamination	2.91 ± 0.89	slight enrichment	2.00 ± 0.38	tends to highly contaminated
Rb	42.95	very severe contamination	6.56	high or very high contamination	8.20 ± 6.95	reasonably high enrichment	0.57 ± 0.45	tends to moderately contaminate
Sr	1.48	reasonable contamination	0.76	low contamination	0.10 ± 0.05	no enrichment	1.82 ± 0.96	moderately contaminated

Calculated DC = 36.66 (high degree of pollution)
Calculated PLI = 1.26 (indicates presence of pollution)

3.1. Analysis and Indices Evaluation

Eco-environment and human health risks are being examined by heavy metal pollution index methods. These indices like DC, CF, EF, I_{geo}, PCI, PERI, and PLI play a vital role in knowing the overall risk with respect to individual heavy metals. The descriptive statistics (mean ± SD) of soil elemental concentration for different altitudes and background concentrations (BCs) of elements are shown in Table 1. Eco-environmental indices with estimated contamination levels in the studied region are provided in Table 4 and a further ecological risk index (Table 5) has been also evaluated.

Table 1 reveals that the elevated concentration of Cu, Cr, Rb, and Zn is chiefly accountable for the soil pollution by an excess amount of heavy metals in the studied region of Uttarakhand Himalaya. The other elements such as Ca and S significantly contribute to soil contamination in some samples. The Al, Ca, Si, Mg, and Na were found to be less than their BC value. Moreover, Cu, Cr, Rb, and Zn were estimated to be more than their BC value. The maximum and mean values of Fe, Ni, and P were estimated at more than their BC value.

Table 5. Calculated potential ecological risk index.

S. No	Elements	Potential Ecological Risk Factor (Pn)	Ecological Risk
1.	Cr	10.69	Low
2.	Mn	2.17	Low
3.	Ni	28.38	Low
4.	Cu	31.95	Low
5.	Zn	2.37	Low
Net PERI		75.56	Low

The higher concentration of elements is an indication of the presence of contamination in the studied catchments and further risk to ecological life; contamination levels were classified based on the calculated values of the indices (DC, CF, EF, I_{geo}, PCI, PERI, and PLI). Further, a comparative analysis was done with an average estimated concentration of elements like Cr, Cu, Ni, Sr, and Zn with their threshold concentration in the soil sample of 100, 30, 80, 200, and 300 mg/kg, respectively [58,86,87]. The estimated average concentration of Cr and Cu was found higher than their corresponding threshold limit; however, Zn was found within its threshold limit. Moreover, mean concentrations of Sr and Ni were estimated more than their respective threshold limit. Geological indices like I_{geo} were estimated for individual elements as per the respective BC. The results reveal that the mean I_{geo} for elements such as Al, Ca, K, Mg, Na, P, S, and Si were found as $I_{geo} \leq 0$, indicating noncontamination of the studied catchment, which signifies no contamination with these elements, whereas elements like Cr, Cu, Fe, Mn, Ni, Rb, and Sr fall under moderately contaminated ($0 < I_{geo} \leq 1$) and/or tends toward moderately contaminated ($1 < I_{geo} \leq 2$) except Zn (which tends toward high contamination, i.e., $2 < I_{geo} \leq 3$). Overall, the I_{geo} analysis reveals that Ni and Rb to some extent, and Cu, Mn, and Cr have a considerable impact on ecological life and further risk to the health of the native population of the studied region of Uttarakhand Himalaya if these elements come into the food chain.

Therefore, regular monitoring of these metals in soils, food products, and vegetables are important to understand the potential health risk and further prevent accumulation of toxic metals in the food chain. Similarly, the EF of each element for the entire study was calculated and normalized against aluminum (Al) to a conservative geogenic element. The mean EF of elements Ca, Mg, Na, and Sr was estimated as 0.42, 0.68, 0.10, and 0.10, respectively, and no enrichment was found of these elements in the studied catchment, whereas Rb, Cu, Ni, and Cr fall under reasonably high enrichment (Rb > Cu > Ni > Cr). Al, Fe, K, Mn, P, S, and Zn that fall into the slight enrichment range ($1 < EF < 3$) might be an indication of the anthropogenic hindrance (e.g., construction of dams and roads) in the vicinity of the studied catchment.

The CF of each element was calculated, and the result reveals that Na, Mg, Si, Al, K, S, Sr, and Ca were found to be less than one (i.e., CF < 1), indicating no contamination in the soil of the studied

Himalayas catchment, while P, Mn, Fe, and Zn showed reasonable contamination (1 < CF > 3) in the studied samples. These elements are generally present in the natural ecosystem and also added to the food producing system as fertilizers, etc., and are beneficial for the growth and development at tracer levels [88]. Different species including humans in healthy conditions have been reported to have the intrinsic capacity to remove the excess elements without much damage. However, continuous exposure to certain elements either through occupation or through contaminated intake along with food, certain medical conditions, impairment of biochemical processes, and accidental intake could be problematic [3,89]. For example, neurotoxicity due to inhalation exposure to airborne Mn has been reported [24,90,91]. Fe toxicity has long been already established by Reissmann et al. [92]. The Zn is an essential element for many biochemically important enzymes in plants as well as animal systems. The permission limit of Zn in Indian soil is 600 µg/g [93]. Over the permissible limits of these elements, they can be problematic/toxic to the living system through impairment of the various physiological, biochemical, growth, and development processes [3,94,95]. Moreover, Cr, Cu, Ni, and Rb fall under considerable contamination to high contamination levels, following the order Rb > Cu > Ni > Cr. Heavy metal toxicity involves oxidative and genotoxic mechanisms [96]. When the direct exposure of toxic elements is not present, these elements can have poisonous effects on the population of a particular region due to accumulation in the food chain [97]. The rubidium (Rb) is observed to be moderately toxic to human as it mimics potassium [26].

Further, PLI was calculated by the conversion factors of each element, and the estimated PLI was found to be 1.26, indicating the presence of pollution in the studied catchment. Calculated PLI results were verified by the degree of contamination (DC) index (Table 4), with results showing a high degree of pollution (DC: 36.66; i.e., DC > 32) at the study site. This high value of DC is due to high anthropogenic activity in the studied catchment. The potential of contamination was calculated using PCI and the results reveal that the studied catchment has low contamination of Na and Si and reasonable contamination of Mg, Al, S, K, Ca, and Sr, whereas other parameters fall between severe to very severe contamination and follow the order Rb > Cu > Mn > Zn > P > Fe. After ascertaining the pollution contamination in the studied catchment, PERI was calculated (using CF and Tn values) to determine the potential ecological risk in the studied catchment (Table 5). The Pn of each element (Cr, Mn, Ni, Cu, Zn) was <40, where PERI was calculated as 75.56, indicating low ecological risk where Ni and Cu contribute a greater share compared to Cr, Mn, and Zn.

3.2. Health Implications of Soil Composition

The topsoil of Uttarakhand Himalaya, India, is well-augmented with Cr, Cu, Mn, Ni, and Zn to the upper crust concentration (UCC), taking into consideration the average BC calculated from the soil distribution values, and above Indian guidelines [98] (Table 6). This soil has Cr and Ni contents above the Indian [98], Canadian [99], and Dutch [100] guidelines for agricultural and construction uses. Cu is above the Canadian guideline for agricultural uses and also above Dutch guidelines (Table 6). The pathways of soil exposure chosen were ingestion, inhalation, and dermal contact, which are considered the same for both (children and adults). Alzheimer's and Parkinson's disease (neurodegenerative disorders) are considered due to the enhanced concentration of Cu, Fe, Mn, and Zn in tissue [101]. Manganese has been found to be a causing agent to induce Parkinsonism through environmental exposure; however, the medical assurance of this conclusion is unclear and needs further investigation [102]. Several elements are present in the natural ecosystem; however, their higher contamination/bioaccumulation than the permissible limits described them as being toxic/nontoxic based on the species (plants, animal, and human)-specific tolerance, which further depend on their interaction in the functioning of the living cells of different species. Most of the toxic elements studies are focused on Cu, Zn, Cr, Ni, Cd, Pb, Hg, As, and Se. However, the other metals might be having an important role to define toxic effects independently or in combination with other toxic metals. The mechanisms of the toxicity/residual/combined effects of these nonstudied metals also need to be understood. Therefore, in the present investigation, all the studied elements and their threshold in light of the indices were evaluated for their hazards. However, for a clearer scenario, more studies are needed to investigate the effect of these reported elements in Himalayan mountainous ecosystems.

Table 6. Mean concentration of heavy metals in the studied site; composition of the upper continental crust and permissible limits (in mg kg^{-1}) as per Indian, Dutch, and Canadian soil.

	Mean Values	Concentration in Upper Continental Crust (B) * (mg/kg)	Indian Guidelines Target Values	Dutch Guidelines Target Values	Canadian Guidelines	
					Agricultural or Other Property Use	Parkland/Residential/ Commercial/Community Property Use
Cr	173	**35**	35	100	67	70
Cu	99	**25**	25	36	62	92
Mn	1342	**600**	-	-	-	-
Ni	99	**20**	20	36	37	82
Zn	143	**71**	71	140	290	290

* Bold values indicated in the tables are the guideline values that are below the mean heavy metal content in Himalaya soils.

Geophagism (soil ingestion), dermal contact, and inhalation are the three key contact pathways of the human health risk from potentially toxic metals and/or elements. Geophagism is very common in children and rarely observed in few adults, while inhalation is related to the dusty composition of air and the dermal contact related to the play and profession in the near-surface environment [9,10,40,41,52]. Table 7 shows the hazard quotient (HQ) values for various pathways and elements resulting from exposure to soil elements.

Table 7. Hazard quotient (HQ) values for various pathways and elements resulting from exposure to soil elements, Uttarakhand, India.

	HQ Ingestion		HQ Dermal		HQ Inhalation	
	Children	Adult	Children	Adult	Children	Adult
Cr	$1.10 \times 10^{+00}$	1.18×10^{-01}	3.07×10^{-03}	4.69×10^{-04}	9.20×10^{-04}	5.19×10^{-04}
Ni	2.57×10^{-02}	2.75×10^{-03}	7.19×10^{-05}	1.10×10^{-05}	7.17×10^{-07}	4.04×10^{-07}
Cu	6.25×10^{-01}	6.70×10^{-02}	1.75×10^{-03}	2.67×10^{-04}	8.81×10^{-04}	4.97×10^{-04}
Mn	$1.10 \times 10^{+00}$	1.18×10^{-01}	3.07×10^{-03}	4.69×10^{-04}	1.47×10^{-02}	8.30×10^{-03}
Zn	8.40×10^{-03}	9.00×10^{-04}	2.35×10^{-05}	3.59×10^{-06}	2.35×10^{-07}	1.32×10^{-07}

For all elements, HQ ingestion is always the highest, while HQ inhalation is always the lowest (Figure 3). The noncarcinogenic HIs for all five elements are given in Table 7. For adults, the HIs were always less than 1, whereas for children, they were higher than 1 for Cr and Mn (Table 7). The HI values of these elements are mainly controlled by HQ ingestion, which are also greater than 1 for these two elements.

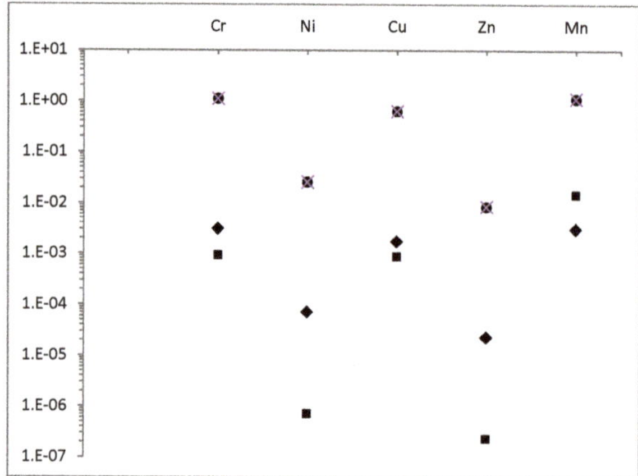

Figure 3. HQs for potentially toxic elements from Uttarakhand, India. Symbols: HQs for ingestion: Crosses; HQs for dermal contact: Diamonds; HQs for inhalation: Squares.

The noncarcinogenic HIs for all five elements are given in Table 8. For adults, the HIs were always less than 1, whereas for children, they were higher than 1 for Cr and Mn. The HI values of these elements are mainly controlled by HQ ingestion, which are also greater than 1 for these two elements. For all elements also, HQ ingestion is always highest, while HQ inhalation is always lowest, except for Mn (Figure 3).

Table 8 shows that the hazard indices (HI) of Cu, Ni, and Zn were below 1; hence, they do not pose a noncarcinogenic risk. For Cr and Mn for children, HI is observed as 1.3 and 4.2, respectively, showing a potential noncarcinogenic risk. The findings of this study were in good agreement with other studies [49]. The case percentage with HI > 1 is 5% and 11% for Cr and Mn, respectively. These findings suggest that children playing should be free from soil exposure, and also, the chance of hand-to-mouth intake should be completely avoided.

Table 8. Maximum and minimum range of hazard indexes and cancer risks due to soil exposure for potentially toxic elements, in Uttarakhand, India.

Element	Hazard Indexes		Cancer Risks	
	Children	Adult	Children	Adult
Cr	0.4–1.3	0.0–0.1	1×10^{-07} to 4×10^{-07}	1×10^{-07} to 1×10^{-06}
Cu	0.0–0.1	0.0–0.0		
Mn	0.2–4.2	0.0–0.6		
Ni	0.0–0.1	0.0–0.0	0 to 5×10^{-09}	0 to 2×10^{-08}
Zn	0.0–0.0	0.0–0.0		

Cr and Ni pose a significant carcinogenic risk as per the International Agency for Research on Cancer (IARC) guidelines [103] in the current investigation. The cancer risk (CR) data for exposure to soil with potentially toxic elements categorized from carcinogenic to possibly carcinogenic to humans [103] indicated that the Cr and Ni cancer risk was up to the standard of carcinogenic risk of 1×10^{-4} to 1×10^{-6} [84] in all the locations of soil sampling.

Indian Himalayan having a rich diversity of flora and fauna, and in recent decades, due to the rapid exploitation of resources, weathering and deforestation, has caused vulnerability to climate change. For the sustainable environmental management and pollution remediation in this region, it is necessary to analyze the concentration of heavy metal and physico-chemical characteristics of the soil together so that its potential risk and its key factor affecting pollution can be determined fruitfully.

The Greater Himalaya includes basic rocks, which are rich in Cr, Ni, Mn, and Cu. Lithosols covering these units are incipient and particularly affected by physical weathering, with their heavy metals contents possibly inherited from parent rocks. Cultivated Cambisols can also explain an anthropogenic origin for contaminants.

All anthropogenic activities underway in the vicinity of the Himalayas with some natural factors are mainly responsible for the soil contamination in the Himalayan region [104,105]. As Uttarakhand Himalaya is vulnerable to climate change, it is highly important to conserve biodiversity, which will further help in socio-economic developments. Preventive measures such as seasonal remediation must be given priority by environmentalists, foresters, and decision-makers to minimize the environmental damages in the coming future. In this study, we suggest that researchers of the different domains conduct a comprehensive investigation on bioavailability, concentration, and transfer of the potential of trace elements in different areas such as agricultural and aquatic in the vicinity of Uttarakhand Himalaya to examine the contamination level and its risk so that its impact on public health and biodiversity (flora and fauna) can be minimized. The environmental standards made by the Ministry of Environment, Forest and Climate Change (MoEF & CC, India) could be helpful for encouraging ecology, and reduced public health risk can enhance the pace of socio-economic developments and limit the human health risks to a great extent. Considering the possible impacts of studied toxic elements on the human health of the native population directly or indirectly through ecosystem consequences, the outcome of this investigation recommends putting forward preventive measures for environmental protection and socio-economic development in the region.

4. Conclusions

This study was conducted in the Uttarakhand Himalaya of India to evaluate the eco-environmental risk and health hazard (carcinogenic or noncarcinogenic risk) to humans using sixteen different soil elements. The elemental concentrations of Cr and Cu were found above their background concentration and threshold limits. The contamination indices (I_{geo}, EF, and CF) evaluated showed that Cu, Cr, and to some extent, Ni and Rb have a considerable impact on soil contamination. The evaluation of DC and PLI obtained values of 36.66 and 1.26, respectively, suggesting that soil of the studied Himalaya catchment is polluted with heavy metals, whereas the calculated PERI was 75.56, indicating low ecological risk. The overall results show no significant ecological risk associated with the heavy metal contents in the soils. Observed elemental contaminations can be ascribed to both geogenic anthropogenic causes and the studied catchment. The health risk assessments for selected heavy metals, whose content was above the Indian, Canadian, and Dutch guidelines, suggest that no major carcinogenic risks for adults were evaluated, due to soil intake, but Cr and Mn concentrations indicate potential carcinogenic risk for children. Therefore, regular monitoring of the reported metals in soils, food products, and vegetables are obligatory to prevent accumulation of metals in the food chain. These key findings will be highly useful for the water resource planners, managers, and environmentalists to make strategic planning in advance to take care of the ecology and human health for people living in the vicinity of the Himalaya catchment.

Author Contributions: Conceptualization and data collection, A.K.[1]; methodology, A.K.[1] and M.C.-P.; formal analysis, A.K.[1] and M.C.-P.; investigation, A.K.[1], M.C.-P. and P.A.D.; resources, A.K.[1] and M.C.-P.; writing—original draft preparation, A.K.[1] and M.C.-P.; writing—review and editing, A.K.[1], M.C.-P., M.K., A.K.[3] and P.A.D.; supervision, A.K.[1]; project administration, A.K.[1]; funding acquisition, A.K.[1] and M.C.-P. All authors have read and agreed to the published version of the manuscript.

Funding: This research was partially funded by the Project UID/GEO/04035/2019 (GeoBioTec Research Centre) financed by FCT—Fundaçãopara a Ciência e Tecnologia.

Acknowledgments: All the authors are highly grateful to the Principal Chief Conservator of Forests (PCCF), Uttarakhand, for permission to carry out work in the forest divisions. A.K.[1] is highly thankful to Prof. M.P Sharma (Prof and Ex. Head, AHEC, IIT Roorkee) for encouragement and support for new findings and guidance from time to time. The authors thank anonymous reviewers.

Conflicts of Interest: The authors declare no conflict of interest.

References

1. Bhatia, A.; Singh, S.D.; Kumar, A. Heavy metal contamination of soil, irrigation water and vegetables in peri-urban agricultural areas and markets of Delhi. *Water Environ. Res.* **2015**, *87*, 2027–2034. [CrossRef] [PubMed]
2. Kumar, A.; Mishra, S.; Taxak, A.K.; Pandey, R.; Yu, Z.G. Nature Rejuvenation: Long-term (1989–2016) vs Short-term Memory Approach based Appraisal of Water Quality of the upper part of Ganga River, India. *Environ. Technol. Innov.* **2020**, *20*, 101164. [CrossRef] [PubMed]
3. Kumar, A.; Kumar, A.; Cabral-Pinto, M.M.S.; Chaturvedi, A.K.; Shabnam, A.A.; Subrahmanyam, G.; Mondal, R.; Gupta, D.K.; Malyan, S.K.; Kumar, S.S.; et al. Lead Toxicity: Health Hazards, Influence on Food Chain, and Sustainable Remediation Approaches. *Int. J. Environ. Res. Public Health* **2020**, *17*, 2179. [CrossRef] [PubMed]
4. Vatanpour, N.; Feizy, J.; Talouki, H.H.; Es'haghi, Z.; Scesi, L.; Malvandi, A.M. The high levels of heavy metal accumulation in cultivated rice from the Tajan river basin: Health and ecological risk assessment. *Chemosphere* **2020**, *245*, 125–639. [CrossRef] [PubMed]
5. Rana, K.; Kumar, M.; Kumar, A. Assessment of Annual Shoot Biomass and Carbon Storage Potential of Grewia optiva: An Approach to Combat Climate Change in Garhwal Himalaya. *Water Air Soil Pollut.* **2020**, *231*, 450. [CrossRef]
6. Adimalla, N.; Chen, J.; Qian, H. Spatial characteristics of heavy metal contamination and potential human health risk assessment of urban soils: A case study from an urban region of South India. *Ecotox. Environ. Safe* **2020**, *194*, 110406. [CrossRef]
7. Galan, E.; Fernández-Caliani, J.C.; Gonzalez, I.; Aparicio, P.; Romero, A. Influence of geological setting on geochemical baselines of trace elements in soils. Application to soils of South–West Spain. *J. Geoch. Exp.* **2008**, *98*, 89–106. [CrossRef]
8. Sundaray, S.K.; Nayak, B.B.; Lin, S.; Bhatta, D. Geochemical speciation and risk assessment of heavy metals in the river estuarine sediments—A case study: Mahanadi basin, India. *J. Hazard. Mater.* **2011**, *186*, 1837–1846. [CrossRef]
9. Cabral Pinto, M.M.S.; Marinho-Reis, P.; Almeida, A.; Pinto, E.; Neves, O.; Inácio, M.; Gerardo, B.; Freitas, S.; Simões, M.R.; Dinis, P.A.; et al. Links between Cognitive Status and Trace Element Levels in Hair for an Environmentally Exposed Population: A Case Study in the Surroundings of the Estarreja Industrial Area. *Int. J. Environ. Res. Public Health* **2019**, *16*, 4560. [CrossRef]
10. Cabral-Pinto, M.M.S.; Inácio, M.; Neves, O.; Almeida, A.A.; Pinto, E.; Oliveiros, B.; Ferreira da Silva, E.A. Human Health Risk Assessment Due to Agricultural Activities and Crop Consumption in the Surroundings of an Industrial Area. *Expo. Health* **2019**, 1–12. [CrossRef]
11. Cabral-Pinto, M.M.S.; Silva, E.A.; Silva, M.M.; Melo-Gonçalves, P.; Candeias, C. Environmental risk assessment based on high-resolution spatial maps of potentially toxic elements sampled on stream sediments of Santiago, Cape Verde. *Geosciences* **2014**, *4*, 297–315. [CrossRef]
12. Cabral-Pinto, M.M.S.; Ferreira da Silva, E.A.; Silva, M.M.V.G.; Melo-Gonçalves, P. Heavy metals of Santiago Island (Cape Verde) top soils: Estimated background value maps and environmental risk assessment. *J. Afr. Earth Sci.* **2015**, *101*, 162–176. [CrossRef]
13. Islam, M.S.; Ahmed, M.K.; Habibullah-Al-Mamun, M. Apportionment of heavy metals in soil and vegetables and associated health risks assessment. *Stoch. Environ. Res. Risk Assess.* **2016**, *30*, 365–377. [CrossRef]
14. Jiang, Y.; Chao, S.; Liu, J.; Yang, Y.; Chen, Y.; Zhang, A.; Cao, H. Source apportionment and health risk assessment of heavy metals in soil for a township in Jiangsu Province, China. *Chemosphere* **2017**, *168*, 1658–1668. [CrossRef] [PubMed]
15. Singh, J.; Rawat, K.S.; Kumar, A. Mobility of cadmium in sewage sludge applied soil and its uptake by radish (*Raphanus sativus* L.) and spinach (*Spinacia oleracea* L.). *Int. J. Agric. Food Sci. Technol.* **2013**, *4*, 291–296. Available online: http://www.ripublication.com/ijafst.html (accessed on 26 July 2020).
16. Singh, K.P.; Malik, A.; Sinha, S. Water quality assessment and apportionment of pollution sources of Gomti river (India) using multivariate statistical techniques—A case study. *Analy. Chim. Acta* **2005**, *538*, 355–374. [CrossRef]

17. Jacob, J.M.; Karthik, C.; Saratale, R.G.; Kumar, S.S.; Prabakar, D.; Kadirvelu, K.; Pugazhendhi, A. Biological approaches to tackle heavy metal pollution: A survey of literature. *J. Environ. Manag.* **2018**, *217*, 56–70. [CrossRef] [PubMed]

18. Albering, H.J.; van Leusen, S.M.; Moonen, E.J.C.; Hoogewerff, J.A.; Kleinjans, J.C.S. Human health risk assessment: A case study involving heavy metal soil contamination after the flooding of the river Meuse during the winter of 1993–1994. *Environ. Health Perspect.* **1999**, *107*, 37–43. [CrossRef]

19. Subki, N.S.; Hashim, R.; Muslim, N.Z. Heavy metals analysis of Batik industry wastewater, plant and soil samples: A comparison study of FAAS and HACH colorimeter analytical capabilities. In *From Sources to Solution*; Springer: Singapore, 2013; pp. 285–289. [CrossRef]

20. Paterson, E.; Sanka, M.; Clark, L. Urban soils as pollutant sinks—A case study from Aberdeen, Scotland. *Appl. Geochem.* **1996**, *11*, 129–131. [CrossRef]

21. Kumar, A.; Sharma, M.P. Carbon stock estimation in the catchment of KotliBhel 1A Hydroelectric Reservoir, Uttarakhand, India. *Ecotox. Environ. Saf.* **2016**, *134*, 365–369. [CrossRef]

22. Adimalla, N. Heavy metals pollution assessment and its associated human health risk evaluation of urban soils from Indian cities: A review. *Environ. Geochem. Health* **2020**, *42*, 173–190. [CrossRef]

23. Eid, R.; Nagla, T.T.; Arab, M.T. Greenwood, Iron mediated toxicity and Programmed Cell Death: A review and a re-examination of existing paradigms. *BBA Mol. Cell Res.* **2016**. [CrossRef]

24. Neal, S.L.; Zheng, W. Manganese Toxicity Upon Overexposure: A Decade in Review. *Curr. Environ. Health Rep.* **2015**, *2*, 315–328. [CrossRef]

25. Exley, C. The toxicity of aluminium in humans. *Morphologie* **2016**. [CrossRef] [PubMed]

26. Tuoni, M.; Taddei, M.; Pasquini, C.; Stefanelli, A.; Meschi, M.; Bionda, A.; Dal Pino, B.; Placidi, G.F. Electrolytes, renal function, and rubidium chloride therapy side effects. In *Recurrent Mood Disorders*; Placidi, G.F., Dell'Osso, L., Nisticò, G., Akiskal, H.S., Eds.; Springer: Berlin/Heidelberg, Germany, 1993. [CrossRef]

27. Cohen-Solal, M. Strontium overload and toxicity: Impact on renal osteodystrophy. *Nephrol. Dial. Transplant.* **2002**, *17*, 30–34. [CrossRef]

28. Kabata-Pendias, A.; Pendias, H. *Trace Elements in Soils and Plants*, 2nd ed.; CRC Press, Inc.: Boca Raton, FL, USA, 1992; p. 365.

29. Tang, Z.; Zhang, L.; Huang, Q.; Yang, Y.; Nie, Z.; Cheng, J.; Yang, J.; Wang, Y.; Chai, M. Contamination and risk of heavy metals in soils and sediments from a typical plastic waste recycling area in North China. *Ecotox. Environ. Saf.* **2015**, *122*, 343–351. [CrossRef]

30. Kumar, A.; Kumar, M. Estimation of Biomass and Soil Carbon Stock in the Hydroelectric Catchment of India and its implementation to Climate Change. *J. Sustain. For.* **2020**, *39*. [CrossRef]

31. Chaffee, M.A.; Berry, K.H. Abundance and distribution of selected elements in soils, stream sediments, and selected forage plants from desert tortoise habitats in the Mojave and Colorado deserts, USA. *J. Arid Environ.* **2006**, *67*, 35–87. [CrossRef]

32. Kumar, S.S.; Kadier, A.; Malyan, S.K.; Ahmad, A.; Bishnoi, N.R. Phytoremediation and rhizoremediation: Uptake, mobilization and sequestration of heavy metals by plants. In *Plant-Microbe Interactions in Agro-Ecological Perspective*; Singh, D.P., Singh, H.B., Prabha, R., Eds.; Springer: Singapore, 2017; pp. 367–394. [CrossRef]

33. Sakram, G.; Sundaraiah, R.; Bhoopathi, V.; Saxena, P.R. The impact of agricultural activity on the chemical quality of groundwater, Karanja Vagu Watershed, Medak District, Andhra Pradesh. *Int. J. Adv. Sci. Tech.* **2013**, *6*, 769–786.

34. Kumar, A.; Mishra, S.; Kumar, A.; Singhal, S. Environmental quantification of soil elements in the catchment of hydroelectric reservoirs in India. *Hum. Ecol. Risk. Assess.* **2017**, *23*, 1202–1218. [CrossRef]

35. Kodom, K.; Preko, K.; Boamah, D. X-ray Fluorescence (XRF) Analysis of Soil Heavy Metal Pollution from an Industrial Area in Kumasi, Ghana. *Soil Sed. Cont.* **2012**, *21*, 1006–1021. [CrossRef]

36. Towett, E.K.; Shepherd, K.D.; Cadisch, G. Quantification of total element concentrations in soils using total X-ray fluorescence spectroscopy (TXRF). *Sci. Total Environ.* **2013**, *463*, 374–388. [CrossRef] [PubMed]

37. Minasny, B.; McBratney, A.B. Regression rules as a tool for predicting soil properties from infrared reflectance spectroscopy. *Chemomet. Intel. Lab. Syst.* **2008**, *94*, 72–79. [CrossRef]

38. Cobo, J.G.; Dercon, G.; Yekeye, T.; Chapungu, L.; Kaszere, C.; Murwira, A. Integration of mid-infrared spectroscopy and geostatistics in the assessment of soil spatial variability at landscape level. *Geoderma* **2010**, *158*, 398–411. [CrossRef]

39. Chauhan, M.; Kumar, M.; Kumar, A. Impact of carbon stocks of Anogeissus latifolia on climate change and socio-economic development: A case study of Garhwal Himalaya, India. *Water Air Soil Pollut.* **2020**, *231*, 436. [CrossRef]

40. Sun, Y.; Zhou, Q.; Xie, X.; Liu, R. Spatial, sources and risk assessment of heavy metal contamination of urban soils in typical regions of Shenyang, China. *J. Hazard. Mater.* **2010**, *174*, 455–462. [CrossRef]

41. Xie, Y.; Chen, T.B.; Lei, M.; Yang, J.; Guo, Q.J.; Song, B. Spatial distribution of soil heavy metal pollution estimated by different interpolation methods: Accuracy and uncertainty analysis. *Chemosphere* **2011**, *82*, 468–476. [CrossRef]

42. Su, Y.Z.; Yang, R. Background concentrations of elements in surface soils and their changes as affected by agriculture use in the desert-oasis ecotone in the middle of Heihe River Basin, North-west China. *J. Geochem. Explor.* **2008**, *98*, 57–64. [CrossRef]

43. Horckmans, L.; Swennen, R.; Deckers, J.; Maquil, R. Local background concentrations of trace elements in soils: A case study in the Grand Ducky of Luxembourg. *Catena* **2005**, *59*, 279–304. [CrossRef]

44. Anderson, S.; Odegard, S.; Vogt, R.D.; Seip, H.M. Background levels of heavy metals in Polish forest soils. *Ecol. Eng.* **1994**, *3*, 245–253. [CrossRef]

45. Chen, J.S.; Wei, F.S.; Zheng, C.J.; Wu, Y.Y.; Adriano, D.C. Background concentrations of elements in soils of China. *Water Air Soil Pollut.* **1991**, *57*, 699–712. [CrossRef]

46. Holmgren, G.G.S.; Meyer, M.W.; Chaney, R.L. Cadmium, Pb, Zn, Cu and Ni in agricultural soils of the United States of America. *J. Environ. Qual.* **1993**, *22*, 335–348. [CrossRef]

47. Bradford, G.R.; Chang, A.C.; Page, A.L.; Bakthar, D.; Frampton, J.A.; Wright, H. *Background Concentrations of Trace and Major Elements in California Soils*; Karney Foundation Special Report; University of California: Riverside, CA, USA, 19 March 1996.

48. Ma, L.Q.; Tan, F.; Harris, W. Concentrations and distributions of eleven metals in Florida soils. *J. Environ. Qual.* **1997**, *26*, 769–775. [CrossRef]

49. Salminen, R.; Tarvainen, T. The problem of defining geochemical baselines: A case study of selected elements and geological materials in Finland. *J. Geochem. Explor.* **1997**, *60*, 91–98. [CrossRef]

50. Salminen, R.; De Vos, W.; Tarvainen, T. *Geochemical atlas of Europe*; Geological survey of Finland: Espoo, Finland, 2006.

51. Cabral-Pinto, M.M.S.; Dinis, P.A.; Silva, M.M.; Silva, E.A.F. Sediment generation on a volcanic island with arid tropical climate: A perspective based on geochemical maps of topsoils and stream sediments from Santiago Island, Cape Verde. *Appl. Geochem.* **2016**, *75*, 114–124. [CrossRef]

52. Jia, Z.; Li, S.; Wang, L. Assessment of soil heavy metals for eco-environment and human health in a rapidly urbanization area of the upper Yangtze Basin. *Sci. Rep.* **2018**, *8*, 3256. [CrossRef]

53. Sadhu, K.; Adhikari, K.; Gangopadhyay, A. Assessment of heavy metal contamination of soils in and around open cast mines of raniganj area, India. *Int. J. Environ. Eng. Res.* **2012**, *1*, 77–85.

54. Krishna, A.K.; Govil, P.K. Heavy metal contamination of soil around Pali Industrial Area, Rajasthan, India. *Environ. Geol.* **2004**, *47*, 38–44. [CrossRef]

55. Krishna, A.K.; Govil, P.K. Soil contamination due to heavy metals from an industrial area of Surat, Gujarat, Western India. *Environ. Monit. Assess.* **2007**, *124*, 263–275. [CrossRef]

56. Krishna, A.K.; Govil, P.K. Assessment of heavy metal contamination in soils around Manali industrial area, Chennai, Southern India. *Environ. Geol.* **2008**, *54*, 1465–1472. [CrossRef]

57. Parth, V.; Murthy, N.N.; Saxena, P.R. Assessment of heavy metal contamination in soil around hazardous waste disposal sites in Hyderabad city (India): Natural and anthropogenic implications. *J. Environ. Res. Manag.* **2011**, *2*, 27–34.

58. Okoyeh, E.I.; Murthy, N.N.; Mohan, R.K.; Krishna, K.A. Heavy metal concentration in the soil and sediment of Koturindustrial area Hyderabad, India. *J. Sci. Res. Rep.* **2015**, *6*, 124–132. [CrossRef]

59. Kumar, A.; Sharma, M.P.; Kumar, A. Green House Gas emissions from Hydropower Reservoirs: Policy and Challenges. *Int. J. Renew. Energy Res.* **2016**, *6*, 472–476.

60. Chauhan, J.S.; Gautam, A.S.; Negi, R.S. Natural and Anthropogenic Impacts on Forest Structure: A Case Study of Uttarakhand State. *Open Ecol. J.* **2018**, *11*, 38–46. [CrossRef]

61. Singh, J.; Rawat, K.S.; Kumar, A.; Singh, A. Effect of sewage sludge and bio-fertilizers on physicochemical properties of alluvial soil. *Biochem. Cell. Arch.* **2013**, *13*, 191–202. Available online: http://www.connectjournals.com/bca (accessed on 26 July 2020).

62. Kumar, A.; Sharma, M.P. Estimation of Soil Organic Carbon in the Forest Catchment of two Hydroelectric Reservoirs in Uttarakhand, India. *Hum. Ecol. Risk Assess.* **2016**, *22*, 991–1001. [CrossRef]

63. Kumar, A.; Sharma, M.P. A modeling approach to assess the Greenhouse gas Risk in Koteshwar Hydropower reservoir, India. Hum. *Ecol. Risk Assess.* **2016**, *22*, 1651–1664. [CrossRef]

64. Gansser, A. *Geology of the Himalayas*; Interscience Publishers: London, UK, 1964.

65. Gansser, A. The Ophiolitic Melange, a world-wide problem on Tethyan examples. *Eclogae Geol. Helv.* **1974**, *67*, 479–507.

66. Searle, M.P. Stratigraphy, structure and evolution of the Tibetan-Tethys zone in Zanskar and the Indus suture zone in the Ladakh Himalaya. *R. Soc. Edinb. Trans. Earth Sci.* **1983**, *73*, 205–219. [CrossRef]

67. Gaetani, M.; Garzanti, E. Multicyclic history of the northern India continental margin (northwestern Himalaya). *Am. Assoc. Pet. Geol. Bull.* **1991**, *75*, 1427–1446.

68. Le Fort, P.; Cuney, M.; Deniel, C.; France-Lanord, C.; Sheppard, S.M.F.; Upreti, B.N.; Vidal, P. Crustal generation of the Himalayan leucogranites. *Tectonophysics* **1987**, *134*, 39–57. [CrossRef]

69. Thakur, V.C. Geology of western Himalaya. *Phys. Chem. Earth* **1992**, *19*, 1–355.

70. Ahmad, T.; Harris, N.; Bickle, M.; Chapman, H.; Bunbury, J.; Prince, C. Isotopic constraints on the structural relationships between the Lesser Himalayan Series and the High Himalayan Crystalline Series, Garhwal Himalaya. *Geol. Soc. Am. Bull.* **2000**, *112*, 467–477. [CrossRef]

71. DeCelles, P.G.; Gehrels, G.E.; Quade, J.; Ojha, T.P. Eocene–early Miocene foreland basin development and the history of Himalayan thrusting, western and central Nepal. *Tectonics* **1998**, *17*, 741–765. [CrossRef]

72. Parkash, B.; Sharma, R.P.; Roy, A.K. The Siwalik Group (molasse): Sediments shed by collision of continental plates. *Sediment. Geol.* **1980**, *25*, 127–159. [CrossRef]

73. Najman, Y.; Garzanti, E. Reconstructing early Himalayan tectonic evolution and paleogeography from Tertiary foreland basin sedimentary rocks, northern India. *Geol. Soc. Am. Bull.* **2000**, *112*, 435–449. [CrossRef]

74. Kohn, M.J.; Paul, S.K.; Corrie, S.L. The lower Lesser Himalayan sequence: A Paleoproterozoic arc on the northern margin of the Indian plate. *Geol. Soc. Am. Bull.* **2010**, *122*, 323–335. [CrossRef]

75. Food and Agriculture Organization of the United Nations (FAO). *Guidelines for Soil Description*, 4th ed.; FAO: Rome, Italy, 2006.

76. Mukherjee, S. Introduction to tectonics and structural geology: Indian context. In *Tectonics and Structural Geology: Indian Context*; Springer: Cham, Switzerland, 2019; pp. 1–5.

77. Taylor, S.R.; McLennan, S.M. The Geochemical Evolution of the Continental Crust. *Rev. Geophys.* **1995**, *33*, 241–265. [CrossRef]

78. Muller, G. Index of geo-accumulation in sediments of the Rhine River. *Geo J.* **1969**, *2*, 108–118.

79. Machender, G.; Dhakate, R.; Prasanna, L.; Govil, P.K. Assessment of heavy metal contamination in soils around Balanagar industrial area, Hyderabad, India. *Environ. Earth Sci.* **2011**, *63*, 945–953. [CrossRef]

80. Maanan, M.; Saddik, M.; Maanan, M.; Chaibi, M.; Assobhei, O.; Zourarah, B. Environmental and ecological risk assessment of heavy metals in sediments of Nador lagoon, Morocco. *Ecol. Ind.* **2015**, *48*, 616–626. [CrossRef]

81. Tomlinson, D.C.; Wilson, J.G.; Harris, C.R.; Jeffery, D.W. Problems in the assessment of heavy metals levels in estuaries and the formation of a pollution index. *Helgoland Mar. Sci. Invest.* **1980**, *33*, 566–575. [CrossRef]

82. Hakanson, L. Ecological risk index for aquatic pollution control: A sediment logical approach. *Water Res.* **1980**, *14*, 975–1001. [CrossRef]

83. Davaulter, V.; Rognerud, S. Heavy metal pollution in sediments of the Pasvik River drainage. *Chemosphere* **2001**, *42*, 9–18. [CrossRef]

84. USEPA (United States Environmental Protection Agency). *Exposure Factors Handbook 2011 Edition (Final)*; United States Environmental Protection Agency: Washington, DC, USA, 2011. Available online: http://cfpub.epa.gov/ncea/risk/recordisplay.cfm?deid=236252 (accessed on 18 September 2019).

85. USEPA (United States Environmental Protection Agency). *Risk Assessment Guidance for Superfund: Volume III—Part A, Process for Conducting Probabilistic Risk Assessment*; EPA 540-R-02-002; United States Environmental Protection Agency: Washington, DC, USA, 2001. Available online: https://www.epa.gov/sites/production/files/2015-09/documents/rags3adt_complete.pdf (accessed on 4 September 2019).

86. USDE (U.S. Department of Energy). *The Risk Assessment Information System (RAIS)*; U.S. Department of Energy's Oak Ridge Operations Office: Oak Ridge, TN, USA, 2013. Available online: https://rais.ornl.gov/ (accessed on 4 September 2019).

87. Linzon, S.N.; Temple, P.J.; Pearson, R.G. Sulfur concentrations in plant foliage and related effects. *J. Air Pollut. Control Assoc.* **1979**, *29*, 520–525. [CrossRef]

88. Reissmann, K.R.; Coleman, T.J.; Budai, B.S.; Moriarty, L.R. Acute intestinal iron intoxication. 1. Iron absorption, serum iron, and autopsy findings. *Blood* **1955**, *10*, 35–45. [CrossRef]

89. Banner, J.R.; Tong, T.G. Iron Poisoning. *Pediatric Clin. N. Am.* **1986**, *33*, 393–409. [CrossRef]

90. Racette, B.A.; Criswell, S.R.; Lundin, J.I.; Hobson, A.; Seixas, N.; Kotzbauer, P.T.; Evanoff, B.A.; Perlmutter, J.S.; Zhang, J.; Sheppard, L.; et al. Increased risk of parkinsonism associated with welding exposure. *NeuroToxicology* **2012**, *33*, 1356–1361. [CrossRef]

91. Bowler, R.M.; Gocheva, V.; Harris, M.; Ngo, L.; Abdelouahab, N.; Wilkinson, J.; Doty, R.L.; Park, R.; Roels, H.A. Prospective study on neurotoxic effects in manganese-exposed bridge construction welders. *NeuroToxicology* **2011**, *32*, 596–605. [CrossRef]

92. Reissmann, K.R.; Coleman, T.J. Acute intestinal iron intoxication. II. Metabolic, respiratory, and circulatory effects of absorbed iron salts. *Blood* **1955**, *10*, 46–51. [CrossRef] [PubMed]

93. Bhatnagar, J.P.; Awasthi, S.K. *Prevention of Food Adulteration Act (Act No.37 of 1954) Alongwith Central and State Rules (as Amended for 1999)*; Ashoka Law House: New Delhi, India, 2000.

94. Onakpa, M.M.; Njan, A.A.; Kalu, O.C. A review of heavy metal contamination of food crops in Nigeria. *Ann. Glob. Health* **2018**, *84*, 488–494. [CrossRef] [PubMed]

95. Gupta, N.; Yadav, K.K.; Kumar, V.; Kumar, S.; Chadd, R.; Kumar, A. Trace elements in soil-vegetable interface: Translocation, bioaccumulation, toxicity and amelioration—A review. *Sci. Total Environ.* **2019**, *651*, 2927–2942. [CrossRef] [PubMed]

96. Eutrópio, F.J.; Ramos, A.C.; Folli-Pereira, M.D.S.; Portela, N.D.A.; Santosh, J.B.D.; Conceição, J.M.D.; Bertolazi, A.A. Chapter 22–heavy metal stress and molecular approaches in plants. In *Plant Metal Interaction Emerging Remediation Techniques*; Ahmad, P., Ed.; Elsevier: Amsterdam, The Netherlands, 2016; pp. 531–543.

97. Kumar, S.; Prasad, S.; Yadav, K.K.; Shrivastava, M.; Gupta, N.; Nagar, S.; Bach, Q.V.; Kamyab, H.; Khan, S.A.; Yadav, S.; et al. Hazardous heavy metals contamination of vegetables and food chain: Role of sustainable remediation approaches—A review. *Environ. Res.* **2019**. [CrossRef]

98. Jiao, X.; Teng, Y.; Zhan, Y.; Wu, J.; Lin, X. Soil Heavy Metal Pollution and Risk Assessment in Shenyang Industrial District, Northeast China. *PLoS ONE* **2015**, *10*, 0127736. [CrossRef]

99. Minister of the Environment (Canada). Soil, Ground Water and Sediment Standards for Use under Part XV.1 of the Environmental Protection Act. Available online: http://www.mah.gov.on.ca/AssetFactory.aspx?did=8993 (accessed on 15 April 2020).

100. Ministry of Housing, Spatial Planning and the Environment (VROM). *Circular on Target Values and Intervention Values for Soil Remediation*; The Netherlands Government Gazette, No. 39: Amsterdam, The Netherlands, 2009; Volume 39.

101. Barnham, K.J.; Bush, A.I. Metals in Alzheimer's and Parkinson's diseases. *Curr. Opin. Chem. Biol.* **2008**, *12*, 222–228. [CrossRef] [PubMed]

102. Ahlskog, J.E. *The New Parkinson's Disease*, 2nd ed.; Oxford University Press: New York, NY, USA, 2016; 544p.

103. IARC (International Agency for Research on Cancer). List of Classifications 2017, 1–123. Available online: https:/monographs.iarc.fr/agents-classified-by-the-iarc/ (accessed on 25 March 2020).

104. Khandekar, J.D.; Bhagwat, P.H.; Wasu, M.B. Study of Physico-Chemical Parameters and Presence of Heavy metals in bore well water at Himalaya Vishwa residential area Wardha. *Sci. Revs. Chem. Commun.* **2012**, *2*, 179–182.

105. Sharma, G.; Singh, A.; Singla, S. Quantitative analysis of Physico-Chemical and Heavy Metal Concentration in the Soil of Indian Himalayan Region. *Environ. We. Int. J. Sci. Technol.* **2018**, *13*, 119–136.

Article

Identifying Groundwater Fluoride Source in a Weathered Basement Aquifer in Central Malawi: Human Health and Policy Implications

Marc J. Addison [1],*, Michael O. Rivett [1], Peaches Phiri [2], Prince Mleta [2], Emma Mblame [2], Modesta Banda [2], Oliver Phiri [3], Wilson Lakudzala [3] and Robert M. Kalin [1]

1 Department of Civil & Environmental Engineering, University of Strathclyde, Glasgow G1 1XJ, UK; michael.rivett@strath.ac.uk (M.O.R.); robert.kalin@strath.ac.uk (R.M.K.)

2 The Ministry of Irrigation and Water Development, Lilongwe Headquarters, Private Bag 390, Lilongwe, Malawi; peachesphiri@gmail.com (P.P.); princemleta@gmail.com (P.M.); emmambalame@gmail.com (E.M.); modesta.banda@gmail.com (M.B.)

3 BAWI Consultants, Private Bag 15, Likuni, Lilongwe, Malawi; oliverphiri0@gmail.com (O.P.); wdzala@gmail.com (W.L.)

* Correspondence: marc.addison@strath.ac.uk

Received: 22 June 2020; Accepted: 18 July 2020; Published: 21 July 2020

Abstract: Consumption of groundwater containing fluoride exceeding World Health Organization (WHO) 1.5 mg/L standard leaves people vulnerable to fluorosis: a vulnerability not well characterised in Malawi. To evaluate geogenic fluoride source and concentration, groundwater fluoride and geology was documented in central Malawi where groundwater supplies are mainly sourced from the weathered basement aquifer. Lithological composition was shown as the main control on fluoride occurrence. Augen gneiss of granitic composition posed the greatest geological fluoride risk. The weathered basement aquifer profile was the main factor controlling fluoride distributions. These results and fluoride-lithology statistical analysis allowed the development of a graded map of geological fluoride risk. A direct link to human health risk (dental fluorosis) from geological fluoride was quantified to support science-led policy change for fluoride in rural drinking water in Malawi. Hazard quotient (HQ) values were calculated and assigned to specific water points, depending on user age group; in this case, 74% of children under six were shown to be vulnerable to dental fluorosis. Results are contrary to current standard for fluoride in Malawi groundwater of 6 mg/L, highlighting the need for policy change. Detailed policy recommendations are presented based on the results of this study.

Keywords: fluoride; groundwater; Sustainable Development Goal 6; water quality; rural community water supply; weathered basement aquifer; hydrogeology; policy change

1. Introduction

The United Nations (UN)'s Joint Monitoring Programme (JMP) has classified fluoride as a Sustainable Development Goal 6 (SDG 6) chemical contaminant of concern for Water and Sanitation [1]. Whilst fluoride in small doses (0.5–1.5 mg/L) is beneficial, above this range the risks of worsening fluorosis conditions, dental–skeletal–crippling, increase [2]. Globally 200 million people may be at risk from fluoride consumption exceeding the 1.5 mg/L World Health Organization (WHO) standard [3,4]. During the Millennium Development Goal (MDG) period and continuing into the current SDG phase, community hand-pumped groundwater supplies have proliferated in low-income, developing countries. This has often occurred without systematic analysis of geo-hazards [5]. Research to investigate groundwater fluoride and associated health risks is required in such countries, including

Malawi, where most rural communities fully rely on groundwater for their drinking-water supply [6]. Groundwater fluoride concentration data may often be sparse and assessment of health risks from natural, geogenic sources of fluoride is frequently non-existent. Science-based policy interventions to support the SDGs in such cases vitally require integrated consideration of geological, geographical, hydrochemical, and water-resource management factors together with risk analysis of human exposure to support regulatory control and provision of safe water supply.

Our recent review of groundwater fluoride in Malawi [7] confirms risks are much less well characterised than other fluoride 'hotspot' countries on the east African rift system (EARS), such as Kenya, Tanzania, and Ethiopia [7–11]. Systematic determination of fluoride risk has not occurred in this EARS periphery location and is despite some documented dental fluorosis in Malawi coinciding with increased groundwater fluoride [12–14]. Our interest in the Nathenje area, central Malawi was triggered by not only observations of dental fluorosis coinciding with somewhat elevated groundwater fluoride (<0.5–7.0 mg/L) [12,13], but also, our perception this fluoride occurrence appeared anomalous and perhaps unexpected given its plateau location removed from the main rift valley and a geological setting dominated by meta-sediments. Assessment of fluoride risk nationally requires such areas of less obvious risk (compared to say deep source hot springs) to be included. Factors controlling intermediate fluoride concentrations occurring mostly below the current (generous) Malawi drinking water standard of 6 mg/L, may well exceed the current WHO 1.5 mg/L standard. This concern motivates our study of the chosen area to inform the national strategy.

The nature of the geology present is expected to be a critical factor. Groundwater fluoride arises from two main geogenic source types: deep sourced hydrothermal inputs and/or shallow rock weathering [3,15,16]. Hydrothermal systems may contain elevated groundwater fluoride, occasionally exceeding 1000 mg/L [3]. Weathering of shallow rocks also contributes where groundwater flows through units rich in fluoride-bearing minerals. Alkaline igneous rocks are an important source of such minerals and globally recognised as a dominant groundwater fluoride source [2,17–19]. Fluoride is preferentially weathered from amphiboles (hornblende) and micas (biotite and muscovite) into groundwater. Apatites containing substituted fluoride are more soluble and an important source [3]. Example cases are reported in India, Ghana, Sri Lanka, and the USA [2,3]. In Sri Lanka, for instance, elevated fluoride occurs where granitic or biotite gneiss lithologies dominate and the nature of the underlying basement rock is the controlling factor on concentrations, with deeper boreholes generally containing higher fluoride [15]. Fluoride concentrations in crystalline basement rocks generally span <1 to 10 mg/L and hence may or may not pose problems [3]. Overall, the preponderance of international evidence confirms significant geological control on groundwater fluoride occurrence. This relationship provides the basis for our geological-based methodology. Whilst anthropogenic sources of fluoride exist, including industrial [20,21] and untreated sanitation [22] sources, these are currently insignificant in our rural study area and most of rural Malawi. Hence, an assumption of geological control on groundwater fluoride occurrence generally is reasonable.

Hence, our hypothesis is that the extent of fluoride occurrence in the weathered basement aquifers of Malawi is geologically controlled. Meaning, fluoride concentrations in groundwater predominantly reflect in-situ local geological composition. We test this hypothesis and investigate the link between fluoride occurrence and human health by integrating our results with health proxy indicators (dental fluorosis). Testing this hypothesis is pivotal to our overarching aim to provide an evidence-based framework that informs science-led policy review of groundwater fluoride risk management currently underway by the Government of Malawi, Ministry of Irrigation and Water Development (MoIWD). The ambition (and current Government planning strategy) is to mitigate risk to human health from groundwater fluoride by incrementally reducing the current Malawian drinking water standard for fluoride [23]. The framework to achieve this ambition is developed and demonstrated herein and comprises:

- Use of borehole water quality surveys to assess groundwater fluoride occurrence;
- Assessment of the hypothesised geological control on observed groundwater fluoride;

- Use of Government of Malawi survey results to assess health risks via proxy dental fluorosis indicators;
- Integrating these lines of evidence to investigate linkages between groundwater fluoride and health and develop risk factors for water points.

The study is the first in Malawi to integrate groundwater fluoride occurrence data with fluorosis health effects attributed to groundwater supply and consumption. The concurrent analysis of water source risks, geological sources and review of policy is multi-faceted and represents a paradigm shift in Malawi's approach in assessing fluoride problems by critically assessing the linkage between groundwater fluoride occurrence, health effect risks and proxy manifestation.

2. Study Area

2.1. Setting

Malawi is situated at the southern periphery of the EARS (Figure 1). In central and northern Malawi, it features as a deep freshwater lake, Lake Malawi [24]. In southern Malawi, the valley floor is exposed at the surface as a series of sedimentary basins [7]. The traditional authority (TA) Mazengera study area is in the district of Lilongwe, central Malawi (Figure 1). It lies on the eastern edge of the Kasungu-Lilongwe Plain, an elevated plateau composed of Lower Palaeozoic–Precambrian basement rock and colluvium of around 6000 km^2 with elevations varying from 1000–1800 m above sea-level (masl). Drainage in the study area generally flows northwest from the Nkhoma and Chilenje hills towards the Lilongwe River, and southeast towards the Linthipe River (Figure 1).

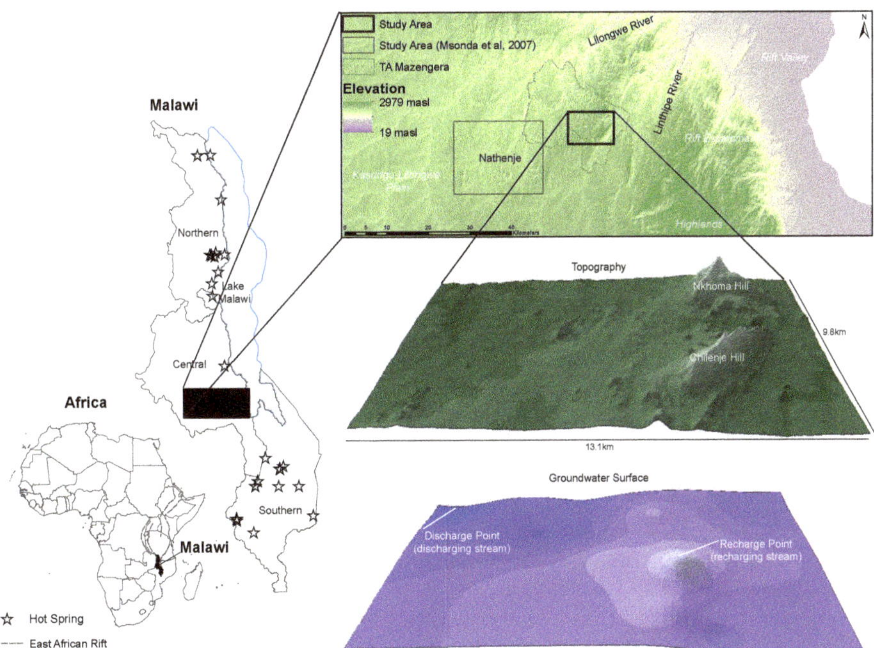

Figure 1. Location of study area within traditional authority (TA) Mazengera alongside Msonda et al. [13] study area at Nathenje area. The figure depicts Malawi's location within the east African rift system, marks the locations of all known hot springs and provides 3-D visualisation of the study area topography and drainage (2× vertical exaggeration), above a 3-D visualisation of groundwater surface indicating the main recharge and discharge points (5× vertical exaggeration).

Climate in the region is sub-tropical/semi-arid with distinct wet (November–April) and dry (May–October) seasons [25]. Monthly rainfall at the peak of the wet season (January) can reach an average of 222 mm, but falls to <1 mm at the dry season height (August). Mean annual precipitation averages 883 mm [26]. This type of climate is conducive to shallow groundwater fluoride enrichment that is found in a similar environment in northeast Sri Lanka. There, the semi-arid climate appears the main factor controlling enrichment due to a number of processes including low precipitation (recharge) and evaporative concentration [15]. Under semi-arid conditions, evapotranspiration at the surface where the water table is shallow may increase fluoride concentrations by a factor of 10–100 [3].

Fluoride occurrence in the study area had not been well characterised prior to this study. The Malawi Government internally released a 'Hydrogeology Atlas (2018)' containing six data points displayed simply as <6 mg/L fluoride [27]. Only one study has conducted fluoride research nearby [13] (Figure 1) which documented fluoride occurrence in the groundwater of Nathenje area, about 20 km southwest of our study area (Figure 1). They sampled 176 boreholes over two years (2001–2002) during both wet and dry seasons. Basement geology comprised an identical range of gneiss lithologies to our study area, however, their western section has considerable colluvium (superficial sediment) which is mostly absent in our case. Over 50% of their samples had fluoride concentrations exceeding the WHO 1.5 mg/L drinking water standard. Concentrations were slightly higher during the dry season, but with spatial trends of highs and lows in concentration comparable across the seasons. Concentrations increased east-northeast leading them to hypothesise high fluoride may be anticipated out with their study area in a geological trend in that direction, towards TA Mazengera, but do not elaborate on specific lithologies. They concluded that in-situ weathering of fluoride-bearing minerals in the basement rock (biotite and hornblende in weathered gneiss) is the probable contributing source of fluoride to groundwater.

Highest groundwater fluoride is typically associated with the rift valley hydrothermal processes and alkaline igneous intrusions [7,28]. Low Ca^{2+} groundwaters were identified as target waters for elevated fluoride (Na-HCO_3, Na-Cl-types) and the breakdown of biotite and hornblende (constituent phases of the basement gneisses of Malawi) is a probable source of fluoride, as fluoride can replace OH^- groups in the crystal lattices of those minerals [13,28]. Water-bearing minerals such as biotite and muscovite are particularly susceptible to hydroxide-replacement [10].

2.2. Geology and Hydrogeology

The study area is 13.1 km by 9.8 km and is located on a plateau some distance from the main rift valley (Figure 1) and is characterised by a distinct lack of major fault systems (Figure 2). The nearest large fracture is 10 km northeast and the nearest rift fault 30 km in the same direction. It is inferred that rift valley processes may exert minimal influence. Structural separation from the rift valley results in no hot spring activity within 70 km of the study area (Figure 1). The nearest is to the northeast at Chikwidzi on the rift escarpment near a large rift-margin normal-fault. The hot spring also represents the closest recorded >6 mg/L (Malawi standard) fluoride concentration [29].

The area comprises almost entirely weathered basement aquifer units [30]. Lithology is dominated by a mix of basement (meta-igneous and meta-sedimentary) rocks: augen gneiss (meta-granite), perthitic syenites, charnockitic gneiss and granulite, hornblende-biotite gneiss and calc-silicate marble. Gneissic foliations and lineations all strike southwest-northeast and dip 40–44° northwest. Hornblende-biotite gneiss occurs as a plunging syncline within charnockitic gneiss and granulite in the northwest of the study area (Figure 2). Nkhoma and Chilenje hills in the east are perthitic syenites which have been intruded into the host basement causing isolated uplift of marbles and steep topography. Small, isolated dambo wetland of river and lacustrine alluvium occur in the south. The western edge contains colluvium deposits which form part of the eastern edge of the superficial deposits' aquifer of the Kasungu-Lilongwe Plain (Figure 2) [31].

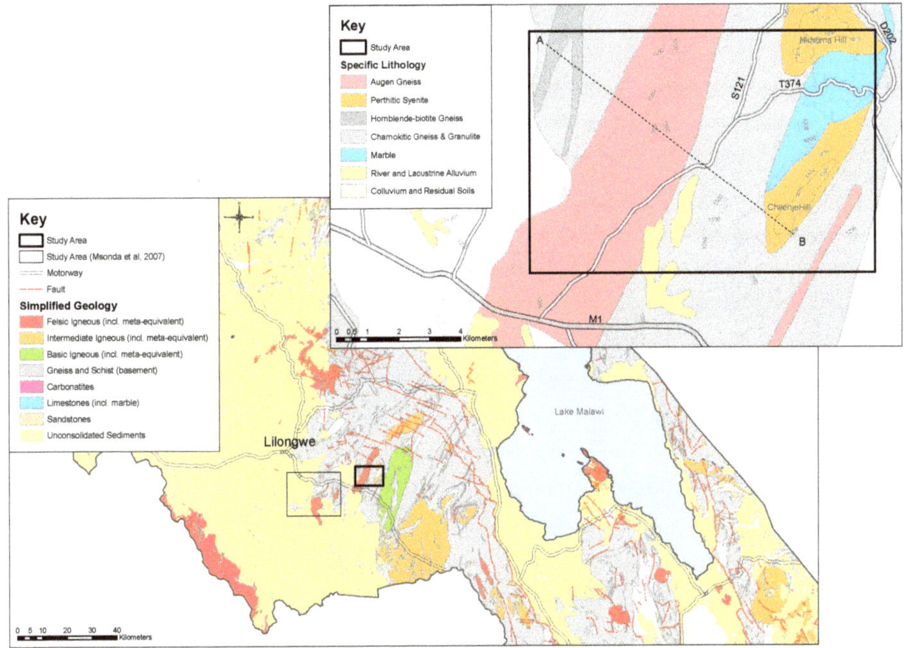

Figure 2. Regional geological map (simplified into lithological type–based on dominant composition) of central Malawi showing relative location of the study area to regional geological features. Map shows structural separation of the study area from the main rift valley (absence of faulting). Inset map shows the specific study area geology and the line of cross-section (A–B) from Figure 3.

Hydrogeological conceptualisation of the wider catchment flow regime from the uplands to Lake Malawi has been described [25]. Catchments originate in the plateau area east of the rift valley and drain northeast towards Lake Malawi, cross cutting the rift escarpment (Figure 1). The study area lies on a small ridge between the catchments for the Lilongwe and Linthipe Rivers and on the divide between catchments, with northwest and southeast flow from the Nkhoma and Chilenje Hills high points. Groundwater levels based on 16 water points (8 boreholes, 4 natural springs and 4 hand dug shallow wells) were used to conceptualise the water table. For the purposes of this conceptual model, only the aquifer west of Chilenje Hill was considered as the hill represents a flow barrier. The primary recharge point is a stream flowing onto an alluvial fan at the base of the northwest slope of Chilenje Hill, natural discharge occurs at a stream in the extreme northwest (Figure 1). Temporal changes in water table elevation (limited by boreholes going dry) suggest seasonal swings may be large at up to 8 m and probably reflect low storage. Geophysical data [32] show that regolith aquifers in this area have uneven thickness, ranging from 0–60 m in depth and are often semi-isolated due to uneven bedrock surface caused by folding. Bedrock often breaches the surface as linear gneiss mounds which strike southwest-northeast, perpendicular to local groundwater flow (assumed from consistent decrease in groundwater elevation southeast-northwest from Chilenje Hill). Pumping tests conducted at five sites within the study area (separate from geochemical sampling sites) show that hydraulic conductivity is variable but low in these aquifers (K = 0.03–0.2 m/day) [33]. A cross-sectional aquifer profile was developed, based on available geophysical and local groundwater data, and is presented in Figure 3a.

The 'weathered basement' conceptual aquifer profile developed (Figure 3b) illustrates the key hydrogeological controls. A regolith aquifer containing saprock and saprolite layers of varying thickness is the main storage unit. Localised fracturing in underlying basement rock increases permeability and limited storage where present, however, the lateral extent of permeability remains unknown. Uneven bedrock surfaces create isolated sub-aquifer units which may become increasingly isolated during the dry season when groundwater levels are low. The aquifer exists under semi-confined conditions due to discontinuous clay layers at the base of the laterite layer. Boreholes drilled where there is clay often have resting water levels above the original water strike [34]. Recharge occurs at the base of the Nkhoma and Chilenje Hills (Figure 1), at alluvial fans, places where unaltered and fractured gneiss is exposed as gneiss mounds and at areas where topsoil/laterite is absent exposing saprolite [32] (Figure 3). Higher groundwater levels may facilitate increased sub-aquifer connectivity (flushing) during the wet season, down-hydraulic-gradient from the Nkhoma and Chilenje Hills in northwest and southeast directions, following decreasing altitude (Figure 3), however this is assumed to be minimal based on low hydraulic conductivities [33] and perpendicular flow barriers.

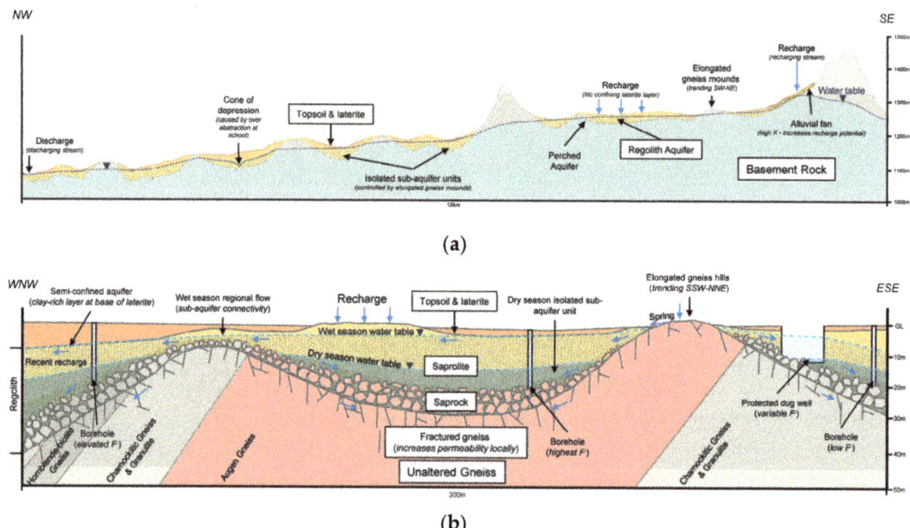

Figure 3. (**a**)—Aquifer cross-section (A–B) of study area showing decreasing altitude and groundwater levels southeast-northwest from recharge at high elevation to discharge at low elevation (vertically exaggerated), topography surface, aquifer depth and water table levels calculated from data. The figure shows semi-isolated nature of regolith aquifer units. (**b**)—Schematic hydrogeological conceptual model (not to scale).

Groundwater resource development is mainly boreholes accounting for 66% of water points (incl. surface water). Drilling may be with little geological or hydrogeological knowledge. Boreholes are drilled to an average depth of 47 m (range: 34–66 m) and cover all lithologies. A smaller proportion of hand dug shallow wells (3–10 m depth) exists (19% of water points), 81% of which are protected at the surface by hand-pumps. The unprotected wells have open sections and are often covered simply by a piece of wood. There are also some natural springs (2% of all water points) at the base of the hills. There is a partially functional piped water supply network in the area (12%) where water is transported from gravity-fed systems and a reticulated borehole, via pipes to various kiosks. Piped supplies were not sampled in this study.

3. Materials and Methods

3.1. Groundwater Survey

Groundwater samples (39) were collected to investigate the geochemical relationship between groundwater and local geology and to test our hypothesised geological control. These covered the three main lithological aquifer sources: augen gneiss, hornblende-biotite gneiss and charnockitic gneiss and granulite. They also covered the range of groundwater supply types: 34 boreholes, three hand dug shallow wells and two natural springs. They comprised 16% of all water points in the area and were evenly distributed. All sampled boreholes and shallow wells were located in regolith and both springs located in bedrock. Samples were collected in November 2019 (wet season) in 1 litre plastic bottles and stored away from light at 4 °C. Purging of boreholes fitted with Afridev hand-pumps was not necessary as all were regularly used prior to sampling, natural springs and shallow wells were sampled in situ. Samples were shipped to Scotland for analysis at the University of Strathclyde. Water temperature was measured at site, using portable measuring equipment. Electrical conductivity (EC), total dissolved solids (TDS) and pH were measured upon delivery at the laboratory using a Mettler Toledo meter (Model MPC 227). All samples were then filtered and anions (F^-, Cl^-, NO_2^-, Br^-, NO_3^-, PO_4^{3-}, SO_4^{2-}) were analysed by ion chromatography (IC Metrohm, 850 Professional) and cations (Ca^{2+}, Mg^{2+}, Na^+, K^+) analysed by a inductively coupled plasma optical emission spectrometer (ICP OES) (iCAP 6200, Thermo Scientific, Waltham, MA, USA). Total alkalinity ($CaCO_3$) was analysed by KONE Aquakem v. 7.2.AQ2 with results multiplied by a factor of 1.22 to estimate total alkalinity (HCO_3^-). In total, 35 of the samples ion-balanced within the conventionally accepted ±5% analysis uncertainty, the remaining four samples within ±10%. The latter were excluded from geochemical analyses but not excluded from fluoride spatial analyses, or analyses where fluoride is the only geochemical component (as fluoride contribution to total anions was <2% in all four samples, therefore, it is unlikely that fluoride would influence the overall ion balance).

3.2. Survey Data

A total of 6804 households in the TA Mazengera study area were visited by Government of Malawi enumerators as part of a wider SDG 6 survey. Dental fluorosis proxy indicator data from this survey was provided to link fluoride occurrence in the area to human health risks. The area reflects a typical low-income, rural population for central Malawi where most of the population gather their daily water from public water points. Only 9.6% of households surveyed had access to a public piped water supply. The survey data was comprised of anonymised questionnaire responses gathered from doorstep interviews with residents on their water use. One question concerned visible symptoms of dental fluorosis.

Survey results provide qualitative proxy information on the visible symptoms of dental fluorosis, but do not constitute a medical or human health survey and results are not definitive in case diagnosis. Results are hence indicative of the condition based on recording of visual symptoms (brown/black staining of the teeth). No inclusion or exclusion criteria were applied. It is recognised that there could be a misappropriation of potential fluorosis respondents who might have had other medically confirmed dental problems, or a history of tobacco (very low in Malawi) or kola use that could yield similar symptoms [35]. Such caveats are recognised.

3.3. Risk Evaluation

3.3.1. Mapping Risk

Demonstration of geological control on groundwater fluoride occurrence is significant as it would permit use of available geological maps to effectively map fluoride risk; i.e., certain lithologies would map as high risk, others intermediate and other low risk. Proving this hypothesis for the study area (ultimately shown herein) allowed a geological fluoride risk map to be developed for the study area based on statistical analysis of groundwater fluoride data with corresponding host lithology. An arbitrary grading system was developed to represent risk of groundwater with fluoride concentrations in excess of the WHO standard (1.5 mg/L) from each lithology in the study area: Grade 0 (unknown risk–no corresponding groundwater fluoride data); Grade 1 (<20% risk); and Grade 2 (>60% risk). Each lithology was represented on the map as a zone with its corresponding risk grade. The map was based solely on the influence of lithological composition to groundwater fluoride concentrations as it assumes very reasonably in the Malawian rural setting that anthropogenic fluoride inputs were insignificant. The map also assumes the local rock mapped has the dominant influence on its groundwater, rather than a neighboring geological unit from which it may have received some inflows. In addition, the evolution of groundwater hydrochemistry along a flow path within the mapped locality does not result in fluoride concentration changes sufficient to alter the host rock risk grading.

3.3.2. Human Health Risk Assessment

To support policy change on fluoride standards in rural drinking water in Malawi, a 'human health risk assessment' was undertaken on each of the water points sampled for geochemistry. The method was based upon the approach introduced by the United States Environmental Protection Agency (USEPA) as a tool to "assess the nature and possibility of adverse health effects in humans who consume highly contaminated water" [36,37]. Similar studies have been conducted since in India and Iran [36,38,39]. The method involves the calculation of a non-carcinogenic risk index, also known as a 'hazard quotient' (HQ) (dimensionless) detailed below [40,41] (in [38]): chronic daily intake (CDI) is calculated (1) [40,41] (in [38]) and defined by the following parameters:

$$CDI = \frac{C * DI * F * ED}{BW * AT} \tag{1}$$

C—Fluoride content in drinking water (mg/L)
DI—Daily Water Intake (l/day)
F—Exposure Frequency (days/year)
ED—Exposure Duration (years)
BW—Average Body Weight (kg)
AT—Averaging Time for non-carcinogens (days)

The HQ is then defined as follows (2) [40,41] (in [38]):

$$HQ = \frac{CDI}{RfD} \tag{2}$$

CDI—Chronic Daily Intake (mg/kg/day)
RfD—Oral Reference Dose (mg/kg/day)

The HQ value indicates the "ratio of the potential exposure to a substance and the level at which no adverse effects are expected as a result of exposure. If HQ > 1, adverse effects are possible" [39]. HQ represents an indication of potential risk from fluoride at any water point. Three risk categories were applied: children (aged 6 years old), children (aged 12 years old) and adults (>19 years old) (adapted from Qasemi et al. [38]). The values used for Equations (1) and (2) were from previous studies [38,39] (Table 1). BW (body weight) and DI (daily water intake) values were based on averages for the age groups chosen and broadly represent the study area population. Once calculated, HQ values were further analysed to explore statistical relationships with geology. This was achieved by comparing calculated proportions of HQ values > 1 for each lithology.

Table 1. Values used to calculate the hazard quotient (HQ) for the study area water points per age group (adapted from: [38,39]). 'C' values are not included as they represent individual fluoride concentrations for each water point and vary from water point to water point.

Risk Exposure Factors	Values for Age Groups			Unit
	Adults (>19 Years Old)	Children (12 Years Old)	Children (6 Years Old)	
C				(mg/L)
DI	2	1.7	1	(l/day)
F	365	365	365	(days/year)
ED	19	12	6	(years)
BW	70	40	15	(kg)
AT	6935	4380	2190	(days)
RfD	0.06	0.06	0.06	(mg/kg/day)

4. Results

4.1. Hydrochemical Observations

Groundwater in the study area is exclusively Ca-Mg-HCO$_3$ type. Hydrochemical data (HCO$_3^-$, Ca^{2+}, Na$^+$, Mg^{2+}, F$^-$ and pH) were plotted with TDS alongside a TDS map (Figure 4). Groundwater is least mineralised around the Nkhoma and Chilenje Hills. Two natural springs were sampled at the base of the southeast slope of Chilenje Hill. These springs contained the least mineralised groundwater. Increasing fluoride concentrations follow a southeast-northwest trend, with the highest fluoride at locations with the highest HCO$_3^-$, Na$^+$ and Mg^{2+}. Low TDS samples from Augen gneiss display notably higher (an order of magnitude) Na$^+$, Mg^{2+} and F$^-$ signatures than low TDS samples from charnockitic gneiss and granulite. Concentrations then increase with TDS following a similar trend in both lithologies. Geochemical data can be viewed in Table S2 in the supplementary materials.

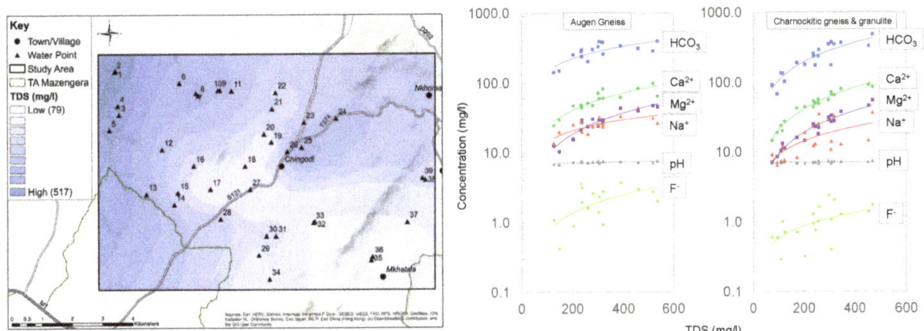

Figure 4. Total dissolved solids (TDS) map of the study area with log plots of relevant hydrochemical concentrations vs. TDS. Map shows TDS values increasing northwest in the direction of decreasing altitude. Graphs show increasing trends in relevant ions with increasing TDS (as seen on the map).

Correlation is not observed between fluoride concentrations and temperature data that span 21–27 °C (groundwater temperatures in Malawi not influenced by geothermal sources are typically around 20–30 °C [7]). Temperatures appear to increase northwest with elevation decline (Figure 1) corresponding to where the water table may be shallow and more vulnerable to surface temperature fluctuations.

Fluoride was plotted with Ca^{2+} alongside fluorite (CaF$_2$) equilibrium (Figure 5). For water with TDS above 200 mg/L, a few water samples approach fluorite saturation but this may be a common ion effect with calcite. Low TDS samples (least mineralised–recent recharge) all plot well below equilibrium displaying both low Ca^{2+} and low fluoride (Figure 5); all are located near the foot of the Nkhoma and Chilenje Hills (Figure 1).

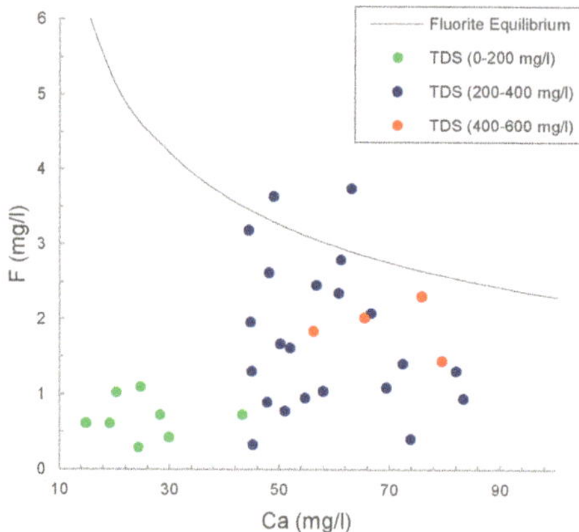

Figure 5. Plot of calcium versus fluoride showing equilibrium for the mineral fluorite (CaF_2). The fluorite equilibrium curve was calculated from $\{F\} = (K_{eq}/\{Ca\})^{0.5}$ that assumes equality of the ion activity product (IAP) with the equilibrium constant (K_{eq}) (3.7×10^{-11}).

A Gibbs diagram plot confirms rock weathering is the dominant process controlling groundwater composition (Figure 6). Just a slight incline towards evaporation suggests minimal evaporation influence on shallow groundwater geochemistry in this system.

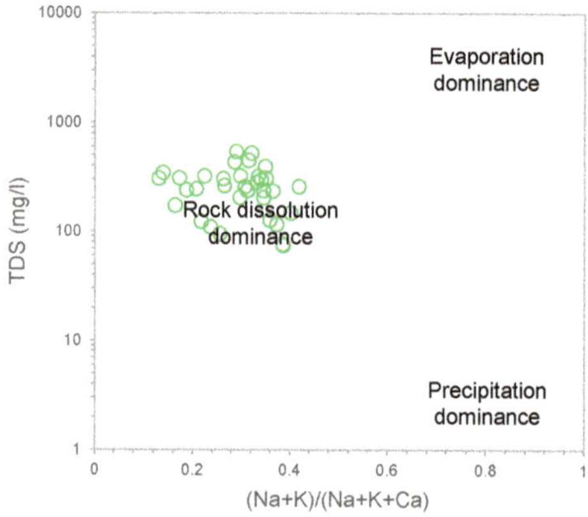

Figure 6. Gibbs diagram of study area groundwater samples showing dominance of rock dissolution. Calculations are shown on axes.

4.2. Geological Controls on Fluoride Occurrence

Fluoride concentrations were lowest in the southeast where the dominant lithology is charnockitic gneiss and granulite, becoming progressively more elevated northwest where the highest fluoride concentrations were located within alkaline meta-igneous rocks (augen gneiss). This lithology contains all fluoride concentrations >3 mg/L. Concentrations then decrease slightly further northwest where the dominant lithology returns to Charnockitic gneiss and granulite, with interbedded hornblende-biotite gneiss (Figure 7). No data were available for perthitic syenite, calc-silicate marble or superficial deposits so their (fluoride) hydrochemical profile remains unknown.

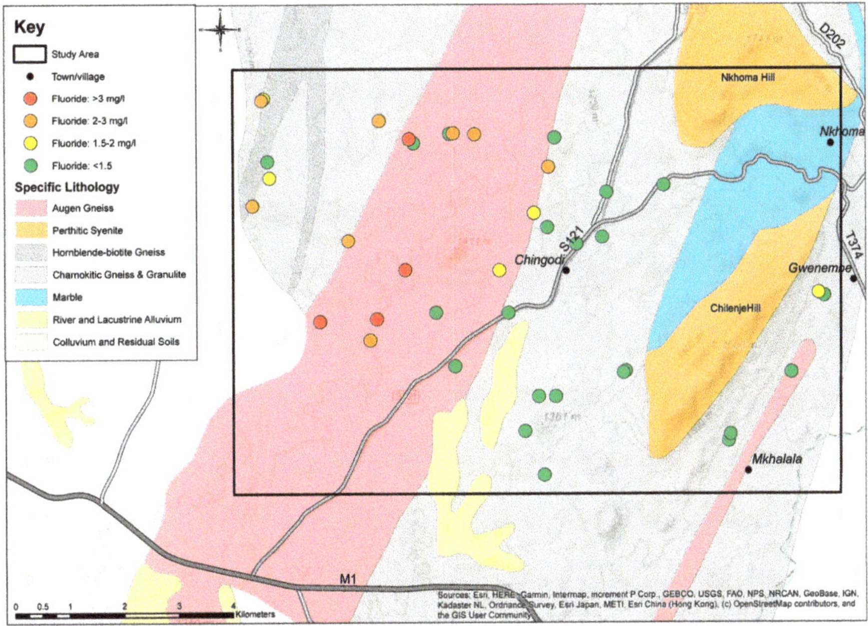

Figure 7. Map showing fluoride data in the study area plotted onto local geology.

A box and whisker plot to investigate the specific lithology-fluoride relationships (Figure 8) shows Charnockitic gneiss and granulite producing the lowest groundwater fluoride with only 20% of samples exceeding the WHO standard for fluoride in drinking water of 1.5 mg/L. A total of 10% of groundwater samples in this lithology exceeded 2 mg/L and none exceeded 3 mg/L. Augen gneiss (alkaline composition) on the other hand had 69% of samples exceeding the WHO standard. This lithology also possesses the highest fluoride concentrations with 56% of samples exceeding 2 mg/L and 25% exceeding 3 mg/L (Table S1—Supplementary Materials). Hornblende-biotite gneiss sits between those lithologies with respect to fluoride concentrations. Low sample numbers for this lithology prevent definitive conclusions, although the expectation would not differ much from those measured values as it contains more fluoride-bearing minerals (relatively) than charnockitic gneiss (biotite and hornblende) and less sodium than the augen gneiss (contains an abundance of Na-plagioclase megacrysts) (Figure 4) which makes it a fluoride source candidate. It has been provisionally included in the (later) risk map for this reason, recognising the need for greater sample numbers.

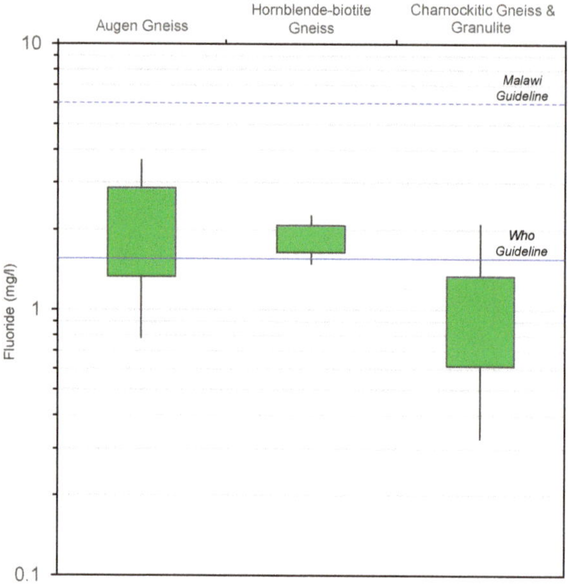

Figure 8. Box and whisker plot of fluoride concentrations (5th, 25th, 75th and 95th percentiles) in the various lithological units (augen gneiss (n = 16), hornblende—biotite Gneiss (n = 3) and charnockitic gneiss and granulite (n = 20)).

4.3. Dental Fluorosis Indicators

Survey responses provide a preliminary indication of dental fluorosis prevalence and are categorised in Table 2 with corresponding host lithologies (only responses within lithologies with groundwater fluoride data are included). Statistically, charnockitic gneiss and granulite had the lowest percentage of "yes" responses at 22% (6% below the average; n = 1022). Augen gneiss hosts the highest percentage of "yes" responses at 41% (13% above the average; n = 853) (Table 2). These results concur with our hydrochemical and geological data, illustrating that groundwater within augen gneiss is the most vulnerable to fluoride, followed by hornblende-biotite gneiss (again, recognising relative low sample numbers for this lithology; n–134) and finally charnockitic gneiss and granulite. This is an important result substantiating our hypothesis.

Table 2. Summary statistics of dental fluorosis indicator data and corresponding geology.

Question Asked	Lithology			Total
"Does anyone in your household suffer from brown-black staining of the teeth?"	Augen gneiss	Hornblende-biotite gneiss	Charnockitic gneiss and granulite	
Total responses	2070	134	4600	6804
"Yes" responses	853	54	1022	1929
% yes	41%	40%	22%	28%

4.4. Risk Evaluation

4.4.1. Risk Map

Statistical analysis of fluoride-lithology data may be hence used to map geological fluoride risk zones (Figure 9). Both augen gneiss and hornblende-biotite gneiss lithologies host >60% risk (68.8%

and 66.7% respectively) of producing drinking water in excess of the WHO standard (1.5 mg/L) and are thus mapped as Grade 2 (highest risk). Statistically, charnockitic gneiss and granulite displayed the least risk (<20%) and were mapped as Grade 1. The blank areas were mapped as Grade 0 as no hydrochemical data were available with which to perform statistical analyses. Red zones (Grade 2), therefore, represent areas where there is >60% risk of abstracted groundwater used for supply producing enough fluoride to cause dental fluorosis (assuming regular consumption from the same water point). Such geology-based risk maps, calibrated to fluoride occurrence, may provide the foundation for the development of risk maps in other areas of similar mapped geologies for which groundwater fluoride data may not exist. With increased groundwater fluoride occurrence data coverage of the presently Grade 0 (unknown) areas it is anticipated greater resolution of the grading system may be possible with increased numbers of grades to characterise the system.

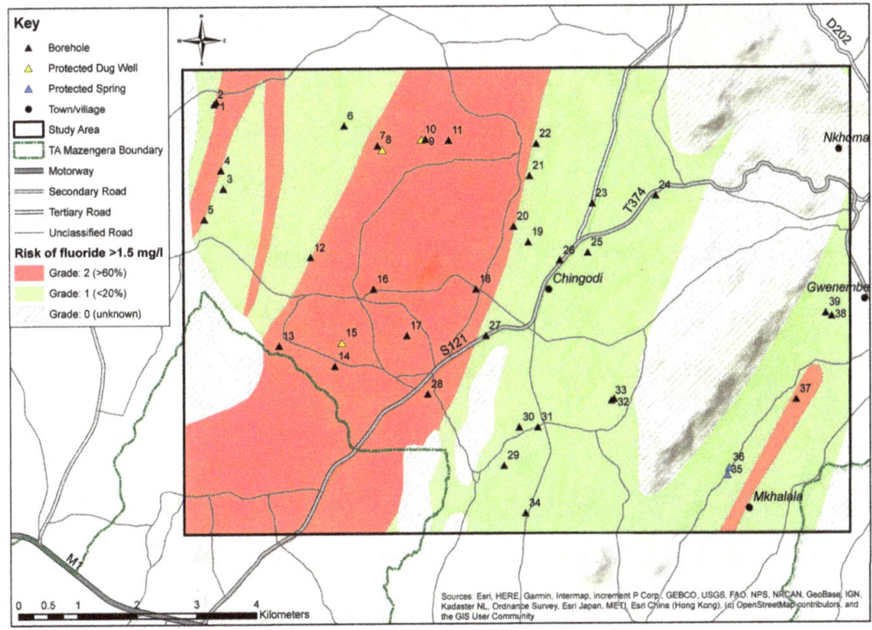

Figure 9. Geological fluoride risk map showing lithologies with risk of groundwater fluoride concentrations >1.5 mg/L as graded zones. Blank zones (Grade 0) reflect areas where geological data are available but corresponding hydrochemical data are not. Sampled water points are shown with their corresponding sample numbers.

4.4.2. Human Health Risk Assessment

74% (*n* = 29) of sampled water points had HQ values >1 for children under the age of 6, indicating possible and increasing risk of dental fluorosis from exposure to fluoride in drinking water from those groundwater points (Table 3). 44% (*n* = 17) of water points had HQ values >1 for children under the age of 12 and 28% (*n* = 11) had HQ values > 1 for adults over 19 years old. The latter also included all children under 12 years old. 26% (*n* = 10) had HQ values <1 for all age groups and appear safe for anyone to drink from. Water Point 13, which displayed the highest HQ value (4.11), also displayed the highest fluoride concentration in the study area (3.75 mg/L) (Table 3).

Table 3. Calculated HQ values for each age group and each water point from the study area. HQ values >1 shown in bold.

Sample Number	Water Point Type	Fluoride (mg/L)	Hazard Quotient (HQ)		
			Adults (>19 years old)	Children (12 years old)	Children (6 years old)
1	Borehole	1.84	0.86	**1.28**	**2.00**
2	Borehole	2.02	0.95	**1.42**	**2.22**
3	Borehole	1.78	0.86	**1.28**	**2.00**
4	Borehole	1.45	0.67	0.99	**1.56**
5	Borehole	2.31	**1.10**	**1.63**	**2.56**
6	Borehole	2.08	**1.00**	**1.49**	**2.33**
7	Borehole	3.04	**1.43**	**2.13**	**3.33**
8	Protected dug well	1.41	0.67	0.99	**1.56**
9	Borehole	2.80	**1.33**	**1.98**	**3.11**
10	Protected dug well	0.89	0.43	0.64	**1.00**
11	Borehole	2.80	**1.33**	**1.98**	**3.11**
12	Borehole	2.46	**1.19**	**1.77**	**2.78**
13	Borehole	3.75	**1.76**	**2.62**	**4.11**
14	Borehole	2.36	**1.14**	**1.70**	**2.67**
15	Protected dug well	3.64	**1.71**	**2.55**	**4.00**
16	Borehole	3.19	**1.52**	**2.27**	**3.56**
17	Borehole	0.43	0.19	0.28	0.44
18	Borehole	1.62	0.76	**1.13**	**1.78**
19	Borehole	1.03	0.48	0.71	**1.11**
20	Borehole	1.96	0.95	**1.42**	**2.22**
21	Borehole	2.62	**1.24**	**1.84**	**2.89**
22	Borehole	1.31	0.62	0.92	**1.44**
23	Borehole	0.78	0.38	0.57	0.89
24	Borehole	1.09	0.52	0.78	**1.22**
25	Borehole	1.42	0.67	0.99	**1.56**
26	Borehole	1.31	0.62	0.92	**1.44**
27	Borehole	0.29	0.14	0.21	0.33
28	Borehole	0.95	0.43	0.64	**1.00**
29	Borehole	0.33	0.14	0.21	0.33
30	Borehole	0.95	0.48	0.71	**1.11**
31	Borehole	0.41	0.19	0.28	0.44
32	Borehole	0.61	0.29	0.43	0.67
33	Borehole	0.73	0.33	0.50	0.78
34	Borehole	0.73	0.33	0.50	0.78
35	Protected spring	0.61	0.29	0.43	0.67
36	Protected spring	0.61	0.29	0.43	0.67
37	Borehole	1.10	0.52	0.78	**1.22**
38	Borehole	1.05	0.48	0.71	**1.11**
39	Borehole	1.67	0.81	**1.20**	**1.89**

Statistical analysis of HQ values with corresponding host lithology (Table 4) demonstrated that for all age groups, augen gneiss carries the most groundwater fluoride risk. For children under the age of 6: 94% of water points in this lithology displayed HQ values >1. Statistically, hornblende-biotite gneiss appeared to be highest risk lithology for that age group (100% of HQ values >1) but again, low sample numbers ($n = 3$) prevents definitive conclusion. The lithology with least fluoride risk was charnockitic gneiss and granulite with only 55% of water points displaying HQ values >1. For children under the age of 12: water points within augen gneiss, again, were the most vulnerable with 69% displaying HQ values > 1. 67% of water points within hornblende-biotite gneiss have HQ values >1 and only 20% for charnockitic gneiss and granulite. The same trend was observed in adults over the age of 19 (Table 4). These results again support augen gneiss as the dominant source of groundwater fluoride, therefore posing the highest dental fluorosis risk.

Table 4. Summary statistics for HQ values >1 per lithology. Values are percentages of sampled water points with an HQ value >1 for each lithology.

Lithology	n	Hazard Quotient (HQ) = > 1		
		Adults (>19 years old)	Children (12 years old)	Children (6 years old)
Augen gneiss	16	50.00%	68.75%	93.75%
Hornblende-biotite gneiss	3	33.33%	66.66%	100.00%
Charnockitic gneiss and granulite	20	10.00%	20.00%	55.00%
All	39	28.21%	43.59%	74.36%

5. Discussion

5.1. Geological Fluoride

Augen gneiss poses the highest potential for elevated groundwater fluoride, ascribed to its granitic-type (alkaline) composition. The augen gneiss is a metamorphosed granite and hosts an abundance of fluoride-bearing minerals such as hornblende, biotite (and accessory apatite) along with characteristic Na-feldspar mega-crysts which are an additional source of sodium (recognised to be conducive to fluoride enrichment). Solubility of fluoride-bearing minerals is generally low. One dimensional reactive-transport equations have shown long residence times are required to produce appreciable concentrations of fluoride in groundwater from silicate rocks, however, higher reactive surface area can significantly increase the rate at which concentrations accumulate [42]; meaning, weathered basement aquifers will have significantly increased fluoride mobilisation potential. Samples represent recent recharge (low mineralisation) and it is likely that increased reactive surface area of the weathered basement aquifer (Saprock, saprolite and fractured gneiss) is a key control on the release of fluoride and may account for observed concentrations. Our geological, geochemical and hydrogeological data support this hypothesis.

Fluorite solubility is usually expected to control fluoride concentrations in groundwater leading to a proportional relationship between F^- and Ca^{2+} ions in solution. Clay minerals and fine sands are moderate adsorbents for fluoride [3], while clay minerals can also be a source of Na^+ for ion exchange with Ca^{2+}, furthering the potential for fluoride enrichment [3,7]. Water type, therefore, plays an important role in fluoride concentration. Lowest fluoride is expected in recharging water, naturally increasing with groundwater evolution through: Ca-$CaCO_3$ (recharge); Ca-Mg-HCO_3; Na-HCO_3; Na-Cl (endmember) [3,28]. While elevated fluoride is expected in geochemically evolved waters, it is commonly found in Ca-Mg-HCO_3-type where there are basement lithologies [39]. This corresponds with our study area where the dominant water type is Ca-Mg-HCO_3 and basement lithologies dominate. Figure 4 shows that Mg^{2+} increases northwest. The loss of Ca^{2+} for Mg^{2+} may facilitate the increasing fluoride trend seen in that direction as there is less Ca^{2+} available to precipitate fluorite (CaF_2). Groundwater samples from augen gneiss plot closest to equilibrium in Figure 5, indicating that equilibration of fluorite (ultimately controlled by equilibration of calcite) is an active process in those samples.

Regional groundwater flow (northwest) is unlikely to be the dominant process producing the hydrochemical trends in that direction (Figure 4). This is due to the shallow, uneven and seasonally isolated nature of the hydrogeological profile (Figure 3) and low hydraulic conductivity values in those aquifers. Local groundwater flow may, however, have increased influence (albeit minimal) during the wet season where there is higher potential for sub-aquifer connectivity via an elevated water table (Figure 3). The surface of unaltered gneiss beneath the study area is locally uneven, with regolith aquifer thickness ranging from 0–60 m below the surface. Elongated gneiss mounds striking southwest-northeast often breach the surface exposing the uneven nature of the bedrock. These may be potential barriers to groundwater flow. The shallow and uneven nature of the regolith controlled by bedrock surface creates (locally) low sub-aquifer connectivity. These results may reflect a slight dry season-wet season shift in groundwater processes where mostly isolated in-situ weathering of basement rock dominates, with the wet season rains allowing a limited degree of sub-aquifer connectivity, flushing groundwater down-hydraulic gradient from the Nkhoma and Chilenje Hills. This slight seasonal shift in processes was assumed from the available data (locally uneven aquifer thickness and a dry season decrease in groundwater levels) as there were no dry season data to compare. Over time, this may allow some down-hydraulic-gradient geochemical groundwater evolution (i.e., transient development of solute concentrations: Figure 4) to occur in that direction, however, the process would be extremely slow and have a relatively insignificant effect on local groundwater hydrochemistry when compared to in-situ weathering. Augen gneiss hosts the lowest hydraulic conductivities (K = 0.003–0.046 m/day) and the outcrop strikes perpendicular to hydraulic gradient, therefore, may act as a boundary to local groundwater flow over lithological boundaries.

Figure 6 confirms rock weathering as the dominant control on fluoride concentrations, with little evaporation or precipitation influence on the hydrochemistry. This supports our hypothesised geological control on groundwater fluoride and significance of in-situ weathering of the underlying geology. Both meta-sedimentary gneiss lithologies play an important role in the concentrations of fluoride that fall within the expected range for rock types [3]. The augen gneiss (alkaline igneous) produced the highest concentrations and is the primary source lithology for fluoride. A geochemical signature for augen gneiss and charnockitic gneiss and granulite appears to be reflected (Figure 4), where low TDS waters (proxy for minimal phreatic influence) from augen gneiss host an order of magnitude higher Na^+ and F^- concentrations than those low TDS waters from charnockitic gneiss and granulite, which may indicate relative parent rock compositional differences (in the absence of rock-powder analyses). These results support the literature consensus that granitic-type rocks host the highest potential for elevated groundwater fluoride [2,15,43]. Augen gneiss outcrops across central Malawi (Figure 2) and may represent a key zone of (dental) fluorosis risk. Hornblende-biotite gneiss on the other hand requires further sampling to be confident in categorising it as a high-risk lithology.

The data support our hypothesis that geology is the dominant control on fluoride occurrence in groundwater in weathered basement aquifers. Weathering of aquifer rock mobilises fluoride into solution, but generally low aquifer transmissivity ensure that fluoride is not transported far from source within the weathered basement profile. This facilitated the development of a map of geological fluoride risk 'zones' (Figure 9). The inherent lack of groundwater fluoride (and other) data for most water points in the study area (and many in Malawi) justifies the need for a preliminary screening approach which can identify high risk zones in the absence of observed groundwater fluoride measurements. User-level risk maps based on Figure 9 could be utilized by local communities to determine if their available, or proposed water points may be at risk from groundwater fluoride. This would be especially useful for users near a zone border who may decide to travel further for water (to a water point within a low-risk zone) to dilute overall fluoride intake. Graded zoning of geological fluoride risk will also prove useful for groundwater development programmes (Government or non-governmental organisations—NGOs) when drilling new boreholes and should be integrated into any subsequent planning strategies. Making informed decisions that avoid drilling in high-risk areas may provide a significant contribution to reducing (dental) fluorosis risk.

5.2. Human Health Risk

Our overarching goal is to inform science-led policy change in Malawi to assist the attainment of SDG targets. For a reduction in drinking water fluoride standard from 6 mg/L to 1.5 mg/L to be achieved, a direct link (risk) to human health from geologically controlled fluoride must be identified and quantified. The calculation of HQ values was completed for each water point sampled. HQ values provide a specific risk factor for each water point, per age group (Table 3) and support other data, identifying augen gneiss as the highest geological fluoride risk lithology (Table 4). HQ values expose the risk from drinking groundwater abstracted directly from high risk lithologies and provide the justification required to advocate policy change. When the HQ values from our sampled water points were compared to the fluoride map contained within the Malawi Hydrogeology Atlas [27], the difference in apparent risk was startling. Malawi's continued use of its 6 mg/L standard implies that the area is safe from dental fluorosis and (ground) water points within are safe to drink (limited to six water points). The HQ values for the same area are contrary, with 74% of water points considered unsafe for children aged 6 years old or under to drink, and 44% of water points unsafe for children under 12 (Table 3).

HQ values have additional value as a means of local risk reduction in the interim. HQ values could be utilised in practice to cycle water use between low and high-risk water points to dilute overall fluoride consumption. For example: anyone over the age of 12 should be safe from developing dental fluorosis by drinking water consistently from water point 30 (F^- = 0.95 mg/L), but a 6-year-old is vulnerable. Water Point 31 (F^- = 0.33 mg/L), however, is safe for anyone to drink (Table 3). By cycling water intake for children aged 6 years and under (also pregnant and breastfeeding women) between Water Points 30 and 31 (50/50), the overall risk for those vulnerable at Water Point 30 is reduced by half, potentially preventing the development of dental fluorosis in those children. If this method can be applied by users at as many of the vulnerable water points as is possible (recognising variable distances between water points), the incidences of dental fluorosis in the study area could be vastly reduced, preventing potentially thousands of people from developing the condition. This method may prove fruitful in the short term via simple planning and informed decision making at user level.

6. Policy and Management Implications with Recommendations

This study began with a challenge: Malawian standard for fluoride in rural (mainly groundwater) drinking water is out of date (currently 6 mg/L) and sufficient understanding of fluoride in over 120,000 rural water supplies must be considered within the SDG 6 timeframe. This research was undertaken to support policy change (Figure 10). A science-based understanding of fluoride occurrence in Malawi combining hydro-geochemical, hydrogeological and human health proxy indicator data was used to quantify the geogenic fluoride risk in a case study area where a weathered basement aquifer dominates groundwater quality. We conducted groundwater surveys to assess groundwater fluoride occurrence, household surveys to assess the extent of the human health impact (dental fluorosis) and water point assessments via geospatial geogenic calculations to quantify the risk to human health from naturally occurring fluoride. The outcomes of this research, specifically the direct health link and potential for geological fluoride risk mapping, has instigated a need within the Malawian Government to review a change in the standard and policy for fluoride in rural water supplies, as scaling of the research outcomes can support new standards in line with WHO. We are now working closely with the MoIWD in Malawi to plan a review and assessment of fluoride risks and implementation of an incremental reduction of the fluoride standard, based on the fluoride risk levels identified (Figure 10).

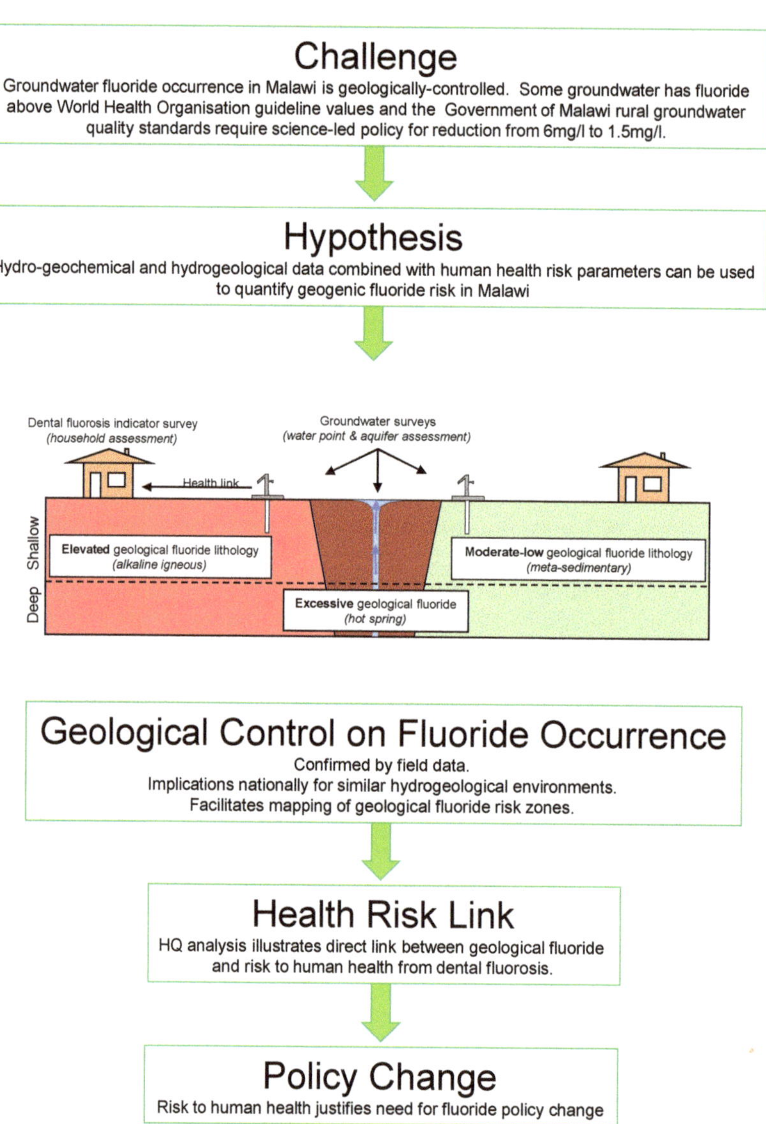

Figure 10. Integrated conceptual model of groundwater fluoride occurrence and health risk linkage leading to advocated groundwater fluoride policy change in Malawi.

This SDG framed research provides tangible and reasonable recommendations which can be implemented within Malawi, including:

- High-level policy change and SDG targets are required for national assessment. Simply changing the fluoride standard from 6 mg/L to 1.5 mg/L is unrealistic and expensive. We propose an incremental decrease in the fluoride standard over time. The 1st stage would be a reduction to 4 mg/L by 2024, instigating an assessment of "excessive fluoride" (hot springs) and "elevated geological fluoride" water points, removing the risk of skeletal fluorosis. Stage 2 would be a reduction to 2 mg/L by 2028, instigating an assessment of "moderate-low geological fluoride" water points and removing the worst of dental fluorosis risk. The final stage would be a reduction to 1.5 mg/L by 2030, bringing their standard in line with the WHO and removing the remaining risk of dental fluorosis from all water points. An evaluation of individual water points in each stage will identify those most harmful and replacement water supplies must be acquired, highlighting the need for incremental change.

- National geological fluoride risk maps should be developed for Malawi. Statistical analyses of fluoride–lithology relationships where fluoride data exist may be translated into risk maps (similar to Figure 9). For areas where fluoride data do not exist, preliminary risk estimates to be later proven may be extrapolated from existing fluoride–lithology data, justified by literature and applied to similar lithologies on a national scale. Risk maps would ultimately be controlled by a synergy of compositional geology and (fluoride) hydrochemistry in non-rift valley zones, and structural geology, compositional geology, hydrothermal processes, and hydrochemistry in rift valley zones. More complex risk models require extensive data sets which Malawi does not currently possess, therefore, mapping geological risk (i.e., fluoride sources) may be the most achievable method of tackling fluoride occurrence at a national scale. National mapping would allow the Government of Malawi and non-governmental organisations (NGOs) active in the water sector in Malawi to integrate fluoride contamination risk into their groundwater resource development strategies.

- Hazard quotient (HQ) values should be shared locally with water point users. It may prove a simple but effective way to inform local people about the potential dangers of each water point and allow them to make informed decisions about water consumption on their own. Decommissioning water points based on elevated fluoride is an expensive venture as a replacement water supply must be provided. In a country with water scarcity problems and low-income, this is a considerable investment planning issue. Revision of rural water quality standards that allow 'yes/no' information on water quality at the water point and information on the negative health effects of fluoride ingestion may prove much more realistic across the country.

- A wider study of varied lithologies should be sampled in the same manner for hydrochemistry to determine their fluoride–lithology relationships. Both perthitic syenite and limestone (marble in this case) lithologies have been linked to high fluoride and are present in the study area but do not currently have corresponding hydrochemical data. National coverage of lithology types is required.

- Collaboration with dental studies in Malawi would be beneficial to corroborating occurrences of fluoride with definitive and documented incidences of dental fluorosis. This will be achieved by working together at the planning stage to ensure both disciplines are conducting their respective research in the same geographical areas. Sharing of data afterwards and working together on cross-discipline publications would ensure the impact of the research to a wider audience of both scientists and policy makers.

- An investigation into piped water supply networks should be undertaken. Piped groundwater from reticulated wells (high-yielding boreholes) drilled to support a network of pipes, powered by solar panels, to numerous kiosks where users can collect groundwater from the same source should be evaluated for fluoride. If a reticulated well is drilled into augen gneiss where fluoride potential concentrations are high, a larger number of people across a wider area will be at risk. The MoIWD or local government (or NGOs) should test for fluoride at kiosks and if found, water cycling with nearby, low fluoride water points should be advised in the first instance. If such methods are not possible at kiosks, those kiosks should be decommissioned, and replacement water supplies installed. If elevated fluoride is found in numerous kiosks fed from the same well, decommissioning of the full system and replacement of the water supply is advised. Future plans for similar piped supplies should incorporate some level of fluoride risk assessment as described by this study. Simply avoiding target (high geological fluoride risk) lithologies may be enough and should be implemented.

Supplementary Materials: The following are available online at http://www.mdpi.com/2076-3417/10/14/5006/s1, Table S1: summary statistics for fluoride data in the study area, Table S2: geochemical data from TA Mazengera samples.

Author Contributions: Conceptualization, M.J.A.; data curation, M.J.A.; formal analysis, M.J.A.; funding acquisition, R.M.K.; investigation, M.J.A.; methodology, M.J.A.; project administration, R.M.K.; resources, M.J.A., O.P. and W.L.; software, M.J.A.; supervision, M.J.A., M.O.R. and R.M.K.; validation, M.J.A., M.O.R., P.P., P.M., E.M. and M.B.; visualization, M.J.A.; writing—original draft, M.J.A.; writing—review and editing, M.J.A., M.O.R., P.P., P.M., E.M., M.B., O.P., W.L. and R.M.K. All authors have read and agreed to the published version of the manuscript.

Funding: This research was funded by The Scottish Government, via the 'Climate Justice Fund: Water Futures Programme', grant number HN-CJF-03 awarded to the University of Strathclyde (R.M. Kalin).

Acknowledgments: The authors gratefully acknowledge the Scottish Government for funding this research. Further thanks are due to Bawi Consultants for their sampling work during Malawi's wet season, without which this work would not have been possible. The sampling strategy was executed with the utmost care and a high level of precision. Finally, the authors would like to thank the Ministry of Irrigation and Water Development, Malawi for their continued cooperation on this project.

Conflicts of Interest: The authors declare no conflicts of interest. The funders had no role in the design of the study; in the collection, analyses, or interpretation of data; in the writing of the manuscript, or in the decision to publish the results.

Ethical Statement: Regarding the use of the Government of Malawi dental fluorosis indicator data, a number of household, functionality, and water quality surveys were carried out in traditional authority Mazengera by Government and private enumerators over the period of the 2019–2020 financial year. Survey responses were anonymous, and all subjects gave their informed consent for inclusion before they participated in the study which complied fully with the Ethical Guidelines of the Malawi Government. All enumerators were professionally trained on the Malawi Government's ethical guidelines which they ably followed in the field during the surveys.

References

1. WHO and UNICEF. *Progress on Drinking Water, Sanitation and Hygiene: 2017 Update and SDG Baselines*; World Health Organization (WHO) and the United Nations Children's Fund (UNICEF): Geneva, Switzerland; New York, NY, USA, 2017; License: CC BY-NC-SA 3.0 IGO.

2. Ozsvath, D.L. Fluoride concentrations in a crystalline bedrock aquifer Marathon County, Wisconsin. *Environ. Geol.* **2006**, *50*, 132–138. [CrossRef]

3. Edmunds, W.M.; Smedley, P.L. Fluoride in Natural Waters. In *Essentials of Medical Geology Revised Edition*; Alloway, B., Centeno, J., Finkelman, R., Fuge, R., Lindh, U., Smedley, P., Selinus, O., Eds.; Springer: Dordrecht, The Netherlands, 2013; pp. 311–350.

4. World Health Organization (WHO). *Guidelines for Drinking-Water Quality: Fourth Edition Incorporating the First Addendum*; World Health Organization: Geneva, Switzerland, 2017; Licence: CC BY-NC-SA 3.0 IGO.

5. Kalin, R.M.; Mwanamvekha, J.; Coulson, A.B.; Robertson, D.J.C.; Rivett, M.O. Stranded Assets as a Key Concept to Guide Investment Strategies for Sustainable Development Goal 6. *Water* **2019**, *11*, 702. [CrossRef]

6. Upton, K.; Ó Dochartaigh, B.; Chunga, B.; Bellwood-Howard, I. Africa groundwater atlas: Hydrogeology of Malawi. *Br. Geol. Surv.* 2018. Available online: http://earthwise.bgs.ac.uk/index.php/Hydrogeology_of_Malawi#:~{}:text=Malawi%20is%20one%20of%20the,settled%20around%20the%2010th%20century (accessed on 1 February 2020).

7. Addison, M.J.; Rivett, M.O.; Robinson, H.; Fraser, A.; Miller, A.V.M.; Phiri, P.; Mleta, P.; Kalin, R.M. Fluoride occurrence in the lower East African Rift System, Southern Malawi. *Sci. Total Environ.* **2020**, *712*, 136260. [CrossRef] [PubMed]

8. Ghiglieri, G.; Pittalis, D.; Cerri, G.; Oggiano, G. Hydrogeology and hydrogeochemistry of an alkaline volcanic area: The NE Mt. Meru slope (East African Rift—Northern Tanzania). *Hydrol. Earth Syst. Sci.* **2012**, *16*, 529–541. [CrossRef]

9. Nair, K.R.; Manji, F.; Gitonga, J.N. The occurrence and distribution of fluoride in groundwaters of Kenya. *Chall. Afr. Hydrol. Water Resour.* **1984**, *61*, 503–512.

10. Olaka, L.A.; Wilke, F.D.H.; Olago, D.O.; Mulch, A.; Musolff, A. Groundwater fluoride enrichment in an active rift setting: Central Kenya. *Sci. Total Environ.* **2016**, *545*, 641–653. [CrossRef] [PubMed]

11. Tekle-Haimanot, R.; Melaku, Z.; Kloos, H.; Reimann, C.; Fantaye, W.; Zerihun, L.; Bjorvatn, K. The geographic distribution of fluoride in surface and groundwater in Ethiopia with an emphasis on the rift valley. *Sci. Total Environ.* **2006**, *367*, 182–190. [CrossRef] [PubMed]

12. Msonda, K.W.M. A Study of Groundwater Quality, Defluoridation and Impact of Dental Fluorosis in Children in Nathenje, Lilongwe, Malawi. Master's Thesis, University of Malawi, Zomba, Malawi, 2003.

13. Msonda, K.W.M.; Masamba, W.R.L.; Fabiano, E. A study of groundwater fluoride occurrence in Nathenje, Lilongwe, Malawi. *Phys. Chem. Earth* **2007**, *32*, 1178–1184. [CrossRef]

14. Sajidu, S.M.; Masumbu, F.F.F.; Fabiano, E.; Ngongondo, C. Drinking water quality and identification of fluoritic areas in Machinga, Malawi. *Malawi J. Sci. Technol.* **2007**, *8*, 42–56.

15. Young, S.M.; Pitawala, A.; Ishiga, H. Factors controlling fluoride contents of groundwater in north-central and northwestern Sri Lanka. *Environ. Earth Sci.* **2011**, *63*, 1333–1342. [CrossRef]

16. Nordstrom, D.K.; Jenne, E.A. Fluorite solubility equilibria in selected geothermal waters. *Geochim. Cosmochim.* **1977**, *41*, 175–188. [CrossRef]

17. Abdelgawad, A.M.; Watanabe, K.; Takeuchi, S.; Mizuno, T. The origin of fluoride-rich groundwater in Mizunami area, Japan—Mineralogy and geochemistry implications. *Eng. Geol.* **2009**, *108*, 76–85. [CrossRef]

18. Naseem, S.; Rafique, T.; Bashir, E.; Bhanger, M.I.; Laghari, A.; Usmani, T.H. Lithological influences on occurrence of high fluoride groundwater in Nagar Parkar area, Thar Desert, Pakistan. *Chemosphere* **2010**, *78*, 1313–1321. [CrossRef] [PubMed]

19. Berger, T.; Mathurin, F.A.; Drake, H.; Astrom, M.E. Fluoride abundance and controls in fresh groundwater in Quaternary deposits and bedrock fractures in an area with fluoride-rich granitoid rocks. *Sci. Total Environ.* **2016**, *569*, 948–960. [CrossRef]

20. Bennett, A.J.; Barratt, R.S. Some observations on atmospheric fluoride concentrations in Stoke-on-Trent. *R. Soc. Health J.* **1980**, *100*, 86–89. [CrossRef]

21. Rivett, M.O.; Ellis, P.A.; Mackay, R. Urban groundwater baseflow influence upon inorganic river-water quality: The River Tame headwaters catchment in the City of Birmingham, UK. *J. Hydrol.* **2011**, *400*, 206–222. [CrossRef]

22. Colombani, N.; Di Guiseppe, D.; Kebede, S.; Mastrocicco, M. Assessment of the anthropogenic fluoride export in Addis Ababa urban environment (Ethiopia). *J. Geochem. Explor.* **2018**, *190*, 390–399. [CrossRef]

23. Malawi Bureau of Standards (MBS). *Drinking Water—Specification*; Malawi Standards Board: Blantyre, Malawi, 2005; ICS 13.030.40 MS 214:2005.

24. Carter, G.S.; Bennett, J.D. *The Geology and Mineral Resources of Malawi*, 2nd ed.; Bulletin No. 6; Government Print: Zomba, Malawi, 1973.

25. Banda, L.C.; Rivett, M.O.; Kalin, R.M.; Zavison, A.S.K.; Phiri, P.; Kelly, L.; Chavula, G.; Kapachika, C.C.; Nkhata, M.; Kamtukule, S.; et al. Water–Isotope Capacity Building and Demonstration in a Developing World Context: Isotopic Baseline and Conceptualization of a Lake Malawi Catchment. *Water* **2019**, *11*, 2600. [CrossRef]

26. Climate-data.org. Available online: https://en.climate-data.org/search/?q=nathenje (accessed on 19 June 2020).

27. Malawi Government. Malawi Government. Malawi hydrogeological atlas, 2018. In *Ministry of Irrigation and Water Development, Malawi*; Government Publication: Lilongwe, Malawi, 2018.

28. Bath, A.H. *Hydrochemistry in Groundwater Development: Report on an Advisory Visit to Malawi. British Geological Survey*; Report No. WD/05/80/20; British Geological Survey: London, UK, 1980; pp. 29–40.

29. Kirkpatrick, I.M. The Thermal Springs of Malawi. *XXIII Int. Geol. Congr.* **1969**, *19*, 111–120.

30. Mkandawire, P.; Chivanga, C. *Hydrogeological Reconnaissance Map: Dedza*, 1st ed.; Department of Surveys; Malawi Government: Blantyre, Malawi, 1987.

31. Thatcher, E.C. *The Geology of the Dedza Area*; Geological Survey Department; Bulletin 29; Government Printer: Lilongwe, Malawi, 1969.

32. Leborgne, R. True 2-D Resistivity Imaging from Vertical Electrical Soundings: Application to Groundwater Investigations in Malawi. Master's Thesis, University of Strathclyde, Glasgow, Scotland, 2019.

33. Storey, I. Determination of Transmissivities through Pumping Tests and their Impact on Water Management, TA Mazengera, Malawi. Master's Thesis, University of Strathclyde, Glasgow, Scotland, 2019.

34. Chilton, P.J.; Smith-Carington, A.K. Characteristics of the weathered basement aquifer in Malawi in relation to rural water supplies. *Chall. Afr. Hydrol. Water Resour.* **1984**, *144*, 57–72.

35. Firempong, C.K.; Nsiah, K.; Awunyo-Vitor, D.; Dongsogo, J. Soluble fluoride levels in drinking water—A major risk factor of dental fluorosis among children in Bongo Community of Ghana. *Ghana Med J.* **2013**, *41*, 16–23.

36. Adimalla, N.; Rajitha, S. Spatial distribution and seasonal variation in fluoride enrichment in groundwater and its associated human health risk assessment in Telangana State, South India. *Hum. Ecol. Risk Assess.* **2018**, *24*, 2119–2132.

37. United States Environment Protection Agency (USEPA): Human Health Risk Assessment. Available online: https://www.epa.gov/risk/human-health-risk-assessment (accessed on 20 December 2019).

38. Qasemi, M.; Afsharnia, M.; Zarei, A.; Farhang, M.; Allahdadi, M. Non-carcinogenic risk assessment to human health due to intake of fluoride in the groundwater in rural areas of Gonabad and Bajestan, Iran: A case study. *Hum. Ecol. Risk Assess.* **2018**, *25*, 1222–1233. [CrossRef]

39. Adimalla, N.; Li, P. Occurrence, health risks, and geochemical mechanisms of fluoride and nitrate in groundwater of the rock-dominant semi-arid region, Telangana State, India. *Hum. Ecol. Risk Assess.* **2018**, *25*, 81–103. [CrossRef]

40. Du, S.; Liu, Y.; Zhang, L.; Li, H.; Huan, H. Assessment of non-carcinogenic health risks due to water contamination in a loess distribution area, northeastern China. *Environ. Earth Sci.* **2017**, *76*, 761. [CrossRef]

41. Yu, R.; Wang, Y.; Ciu, Z.; Xu, G.; Guan, Z.; Yu, Y.; Liu, J. Human health risk assessment of organophosphorus pesticides in maize (*Zea mays* L.) from Yushu, Northeast China. *Hum. Ecol. Risk Assess.* **2017**, *24*, 642–652. [CrossRef]

42. Banerjee, A. Groundwater fluoride contamination: A reappraisal. *Geosci. Front.* **2015**, *6*, 277–284. [CrossRef]

43. Sajil Kumar, P.J.; Jegathambal, P.; Nair, S.; James, E.J. Temperature and pH dependent geochemical modelling of fluoride mobilization in the groundwater of a crystalline aquifer in southern India. *J. Geochem. Explor.* **2015**, *156*, 1–9. [CrossRef]

Article

Retrieval of Turbidity on a Spatio-Temporal Scale Using Landsat 8 SR: A Case Study of the Ramganga River in the Ganges Basin, India

Mona Allam [1,2], Mohd Yawar Ali Khan [3,4,*] and Qingyan Meng [1,*]

[1] State Environment Protection Key Laboratory of Satellite Remote Sensing, Aerospace Information Research Institute, Chinese Academy of Sciences, Beijing 100101, China; monaallam@sci.asu.edu.eg

[2] Environment & Climate Change Research Institute, National Water Research Center, El-Qanater El-Khairiya 13621/5, Egypt

[3] Department of Hydrogeology, King Abdulaziz University, Jeddah 21589, Saudi Arabia

[4] Department of Earth Sciences, Indian Institute of Technology Roorkee, Roorkee 247667, India

* Correspondence: makhan7@kau.edu.sa (M.Y.A.K.); mengqy@radi.ac.cn (Q.M.)

Received: 26 April 2020; Accepted: 20 May 2020; Published: 27 May 2020

Abstract: Nowadays, space-borne imaging spectro-radiometers are exploited for many environmental applications, including water quality monitoring. Turbidity is a standout amongst the essential parameters of water quality that affect productivity. The current study aims to utilize Landsat 8 surface reflectance (L8SR) to retrieve turbidity in the Ramganga River, a tributary of the Ganges River. Samples of river water were collected from 16 different locations on 13 March and 27 November 2014. L8SR images from 6 March and 17 November 2014 were downloaded from the United States Geological Survey (USGS) website. The algorithm to retrieve turbidity is based on the correlation between L8SR reflectance (single and ratio bands) and insitu data. The b2/b4 and b2/b3 bands ratio are proven to be the best predictors of turbidity, with $R^2 = 0.560$ ($p < 0.05$) and $R^2 = 0.726$ ($p < 0.05$) for March and November, respectively. Selected models are validated by comparing the concentrations of predicted and measured turbidity. The results showed that L8SR is a promising tool for monitoring surface water from space, even in relatively narrow river channels, such as the Ramganga River.

Keywords: Ramganga River; turbidity; Landsat 8 SR; water quality; Ganges River

1. Introduction

Turbidity is animportant parameter for water quality and a surrogate for the transparency of water [1–5]. Turbidity can damage many aquatic organisms and fishes by degrading spawning grounds, reducing feed supplies, and affecting gill function [6]. A decrease or increase in water transparency can adversely affect the organic components of systems that adjustto light-dispersing environments [7–15]. In estuarine waters with high turbidity, dissolved oxygen concentrations can significantly decrease due to irregularities in heterotrophic and autotrophic processes, which may contribute to the depletion ofmarine organisms [16,17]. Typically, turbidity is assessed visually using aSecchi disk, or presumably through nephelometry [1,5]. However, these methods only represent the locations from which the sample was collected. Recently, remote sensing of sea color has become a valuable method to retrieve and monitor suspended sediment concentration (SSC) and turbidity in coastal turbid waters on the surface [18–21]. Traditional water quality sampling is cost-effective and time-consuming, as it involves the collection and analysis of the water. Also, the traditional method of water monitoring does not provide the spatial or temporal view of the entire body of water that is necessary for proper management [22].

The use of remote sensing technology to analyze water quality is concise; it alsocaptures the entire field of study to create consistent surface data and periodically demonstrate the point-by-point spatial variability of water quality [23]. Although research on various remote sensing technologies is devoted to total SSC retrieval, research on retrieving turbidity is limited [21]. Even thoughsatellite remote sensing cannot identify near-bed absorption, it is used to identify spatial and temporal variations in turbidity at the surface."Near-bed absorption" refers to the bottom of a stream or other body of water. The turbidity of the surface water can influence the reflection of the water body, but not from the bottom of the stream/lake/ocean, because the reflection data are obtained from the top 2 m of surface water [24].

The Ramganga River is an important tributary of the Ganges River. It originates from the lower Himalayas in Uttarakhand, covering the vast Ganga Flood Plains (GFP) of Uttar Pradesh, and then converges with the Ganges River. It is the primary source of water for the Jim Corbett National Park (Tiger Reserve) situated in the Uttarakhand, and is one of the critical water sources for domestic, industrial, and agricultural use in the western Uttar Pradesh [25,26]. The upper reaches of the study area consist of hillocks and streams, while agricultural fields mainly dominate the middle and lower reaches; therefore, when sufficient rainfall increases the contribution of suspended substances, due to weathering and erosion processes in the upper regions and agricultural runoff in the middle and lower regions, the turbidity and total SSC increase considerably [27]. The aquatic life of the Ramganga River is negatively affected by the large amount of turbidity in the water, and harmful bacteria and pollutants may also be associated with the particles that cause turbidity. Estimating turbidity distribution in the Ramganga River with diverse geomorphology and a complex environment requires anunconventional approach. Remote sensing technology provides reliable information for monitoring and understanding the variation of turbidity in time and space, particularly in the substantial zone with limited access, such as Jim Corbett National Park area of the Ramganga River Basin.

Mapping turbidity and other indicators of water quality is routinely performed using information acquired with wide-swath imagingspectro-radiometers designed to measure sea color—for example, Orbview-2/SeaWiFS, ENVISAT/MERIS, and Aqua/MODIS [28]. However, these applications are not suitable for narrow and small regions, due to their low spatial resolution scales, yielding a large number of mixed pixels and resulting in lower accuracy of retrievals [29]. In comparison to these medium resolution images, Landsat 8 surface reflectance (L8SR) images aredelivered on a Polar Stereo (PS) or universal transverse Mercator (UTM) mapped grid with 30 m spatial resolution.

Table 1 shows the important features of the L8SR product. Various surveys of remote detection of ocean color were carried out to retrieve water quality parameters, most of which used three basic strategies: (i) implicit, based on the correlation between water quality parameters, using inherent optical properties (IOPs) and semi-analytical models [30–32]; (ii) using experimental models between these parameters and IOPs [33,34]; and (iii) experimental models using water quality parameters and satellite data reflection [35–38]. The third approach was used in this study, which is based on the correlation between field measurements and reflectance values extracted from L8SR products.

Table 1. Significant features of Landsat 8 surface reflectance (L8SR) data product.

Band Designation	Band Name	Data Type	Units	Range	Valid Range	Fill Value	Saturate Value	Scale Factor
ProductID_sr_band1	Band1	INT16	Reflectance	−2000–16,000	0–10,000	−9999	20,000	0.0001
ProductID_sr_band1	Band2	INT16	Reflectance	−2000–16,000	0–10,000	−9999	20,000	0.0001
ProductID_sr_band2	Band3	INT16	Reflectance	−2000–16,000	0–10,000	−9999	20,000	0.0001
ProductID_sr_band3	Band4	INT16	Reflectance	−2000–16,000	0–10,000	−9999	20,000	0.0001
ProductID_sr_band4	Band5	INT16	Reflectance	−2000–16,000	0–10,000	−9999	20,000	0.0001
ProductID_sr_band5	Band6	INT16	Reflectance	−2000–16,000	0–10,000	−9999	20,000	0.0001
ProductID_sr_band6	Band7	INT16	Reflectance	−2000–16,000	0–10,000	−9999	20,000	0.0001

2. Materials and Methods

2.1. Study Area

Ramganga River flows through the Himalayas (Kumaon region) in Uttarakhand and the GFP before joining the Ganges River in Uttar Pradesh. The river has a catchment area of approximately 22,685 km^2, with a total stretch of 642 km from its origin (Dudhotali Mountain of the district Chamoli) to the confluence with the Ganges River [39–42]. The Ramganga River catchment lies between 30°06′02.22″ N to 27°10′42.11″ N and 79°16′59.22″ E to 79°50′16″ E, with a mean elevation of 1530 m above mean sea level. After covering the first 158 km of its stretch in the Kumaon Himalayas and going through the Jim Corbett National Park, the river enters the GFP at Kalagarh town, where the Ramganga Dam has been constructed. In the GFP, the river flows through the hugely populated and highly agricultural and industrialized districts of the Uttar Pradesh, such as Moradabad, Bijnor, Bareilly, Rampur, Hardoi, Shahjahanpur, and Farrukhabad [43].

2.1.1. Climatic Condition and Rainfall

Summer, rainy, and winter arethe three distinct seasons witnessed by the study area. The rainy season begins by the middle of June and continues to September or mid-October. Following a brief spell of autumn starting in mid-October, when the temperature drops drastically, the winter season begins in November. October/November and May/June are considered to be thepost-monsoon and pre-monsoon seasons, respectively. Throughout the winter months, some occasional showers also occur (http://indiawaterportal.org/). The average yearly rainfall receives by the area is around 1000 mm [28]. The relationship between water discharge (Q) and the SSC in the Ramganga River are shown in Figure 1. It is clear from the figure that there is a direct relation between Q and SSC.

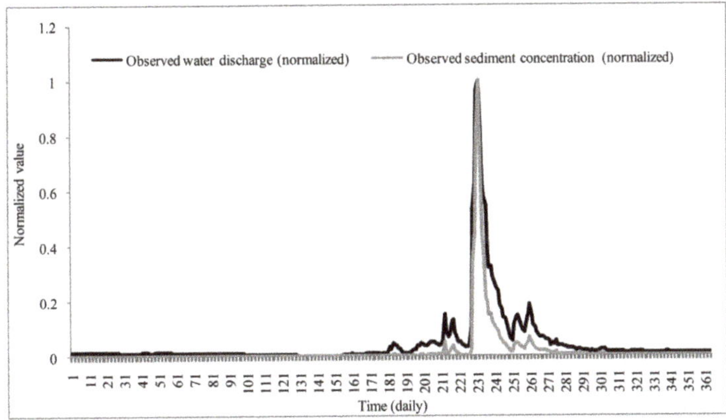

Figure 1. Plot showing relationship between sediment concentration and water discharge in the Ramganga River.

2.1.2. Geology

The entire catchment region is comprised oftwo major lithotectonic zones, namely the Sub-Himalayas and Lesser Himalayas. In the Sub-Himalayas, the major components are siltstone, sandstone, clays, and boulders, with molasse sediments of Mid-Miocene to Pleistocene age. Unfossiliferous sequences of low- to high-grade meta-sediments of the Palaeozoic to Mesozoic age are the major components of the Lesser Himalayas. In general, the important lithologies in the Ramganga basin are (1) calcareous shales and siltstones (Blaini/Infrakrol formations); (2) quartzites (Nagthat and

Sandra formations); (3) low-grade metamorphics (phyllites, slates, and schists); (4) limestones (Krol and Deoband formations); and (5) high-grade metamorphics (granite gneisses) [44].

The river emerges in the Ganga alluvial plain, also known as the GFP, after covering a distance of about 158 km in the Kumaon Himalayas. The Ganga alluvial plain is a foreland basin closely linked with the extension of the Himalaya orogenic belt, as demonstratedin Figure 2. The Quaternary lithostratigraphic sequence established in descending order is comprised of the (1) Ganga/Ramganga Recent Alluvium; (2) Ganga/Ramganga Terrace Alluvium; and (3) Varanasi Older Alluvium, with two facies, i.e., sandy facies and silt clay facies.The first two, the Recent and Terrace alluviums, constitute the Newer Alluvium [45].

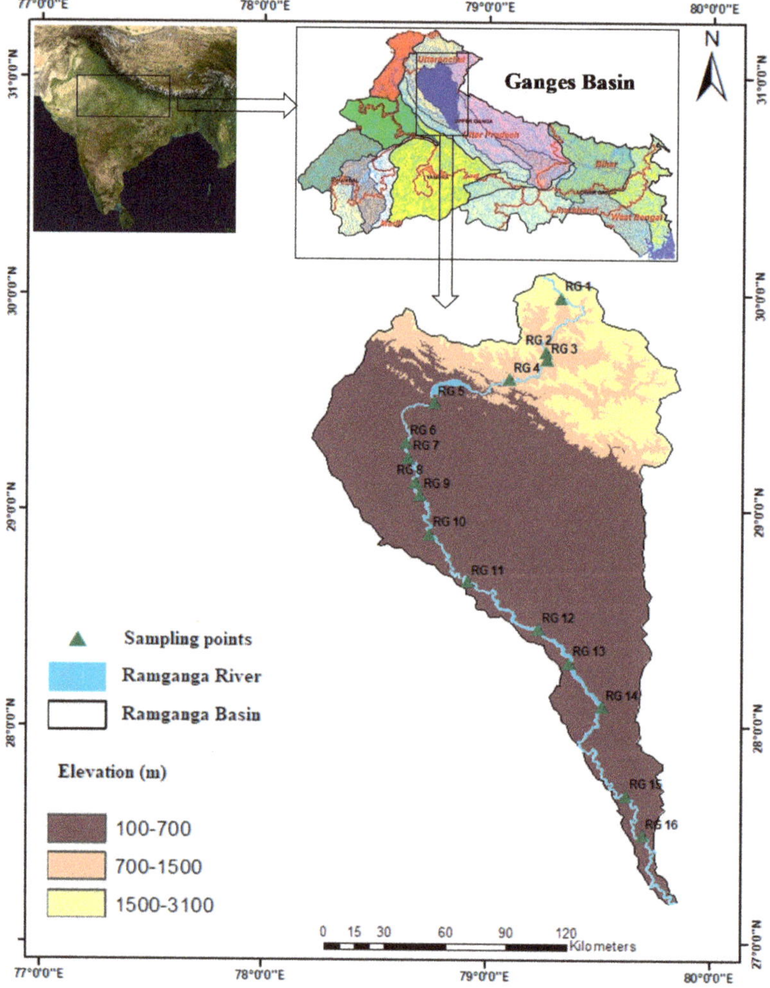

Figure 2. Map showing the Ramganga River and the locations of samples.

2.2. Sample Collection and Analysis

Figure 2 shows the locations where water samples were collected from the Ramganga River. Sampling process was done on 13 March and 27 November of the year 2014 (Table 2). Sixteen samples of river water from each location were collected in a five-liter bottle, preserved, and transferred to the laboratory as suggested in *Standard Methods for the Examination of Water and Wastewater* (APHA), 20th edition [46].

Table 2. Sampling sites description with insitu concentration of turbidity.

Sample ID	Longitude	Latitude	Turbidity (NTU)-March	Turbidity (NTU)-November
RG1	79.321581	29.984017	4.310	0.6
RG2	79.255436	29.732233	5.600	1.2
RG3	79.261153	29.696792	2.820	0.5
RG4	79.093611	29.606047	0.888	0.6
RG5	78.761167	29.496639	5.270	3.5
RG6	78.636108	29.314433	24.600	14.2
RG7	78.649336	29.243347	52.600	8.9
RG8	78.679081	29.127161	20.600	13.3
RG9	78.698394	29.068136	75.900	15.4
RG10	78.744111	28.890639	112.000	2.5
RG11	78.912031	28.668564	99.900	2.3
RG12	79.229528	28.449917	106.000	2.9
RG13	79.368028	28.294722	28.900	3.2
RG14	79.513861	28.094222	41.700	2.7
RG15	79.623308	27.681989	64.500	2.1
RG16	79.697544	27.497983	42.500	2.4

The sample bottles were rinsed with 2% nitric acid in the laboratory, and rinsed twice with water of the river at the time of sampling to avoid contamination. Turbidimeter (HACH instruments) was used to measure the turbidity in NTU of each water sample.

2.3. Satellite Images

It was observed that all the sampling locations occurred in three images (path 145 and rows 139, 140, and 141). The three images cover an area of approximately 180 km east–west to 540 km north–south. There could be significant variability in the atmospheric conditions over such an area, which affects the relationship between the top of the atmospheric reflectance retrieved from the satellite data and the insitu water turbidity. This problem is mitigated by using a single image where 13 of the 16 sampling sites were located. The reflected electromagnetic solar radiation is the basis for the spectral examination of satellite imagery, issued to measure turbidity. Unique signatures and curves are generated, depending on the reflection and absorption at different wavelengths [47,48]. The major errors in the reflected electromagnetic solar radiation remained when retrieving water properties from satellite images. The thirteen samples (RG2–RG14) included in the analysisare located on the image, with path 145 and row 40. Nine samples wereused to retrieve the turbidity model. To validate this model, the measured and predicted turbidity was compared. The four samples that were not included in the model retrieving were used for further validation of the model.

2.4. Statistical Summary of Ramganga River In Situ Measurements

Insitu concentrations of turbidity were measured in both March and November 2014. The distribution of data of turbidity was generally skewed, with low values and without any outliers or very high values (Table 3). Turbidity concentrations ranged between 20.6 and 112.0 NTU with a mean value of 62.467 NTU during March 2014, and between 2.3 and 15.4 NTU with mean value of 7.267 NTU during November 2014.

In general, turbidity concentrations were higher in March than in November (Figures 3 and 4). The SSC depends on the location and time of the year. When matched to the pre-monsoon and post-monsoon data, the SSC values weremuch higher during the monsoon months. Thisis caused by high Q, leading to high rates of weathering and erosion from the catchment and the river channel itself. Pre-monsoon concentrations (March 2014) are consistently higher than the corresponding post-monsoon concentrations (November 2014). Thiscan be attributed to a considerable difference in elevation levels of 530 m (RG4) to 259 m (RG5) from the mean sea level. This elevation difference leads to a decrease in potential energy and an increase in the kinetic energy of the river, thereby increasing the sediment-carrying capacity of the river [16].

Table 3. Descriptive statistics values of turbidity in March and November 2014.

Parameters	March 2014	November 2014
Number of Samples	9	9
Mean	62.47	7.27
Standard Error of the Mean	12.23	1.89
Standard Deviation	36.70	5.67
Variance	1346.55	32.14
Skewness	0.27	0.55
Standard Error of Skewness	0.72	0.72
Kurtosis	−1.88	−1.92
Standard Error of Kurtosis	1.4	1.4
Range	91.4	13.1
Minimum	20.6	2.3
Maximum	112.0	15.4

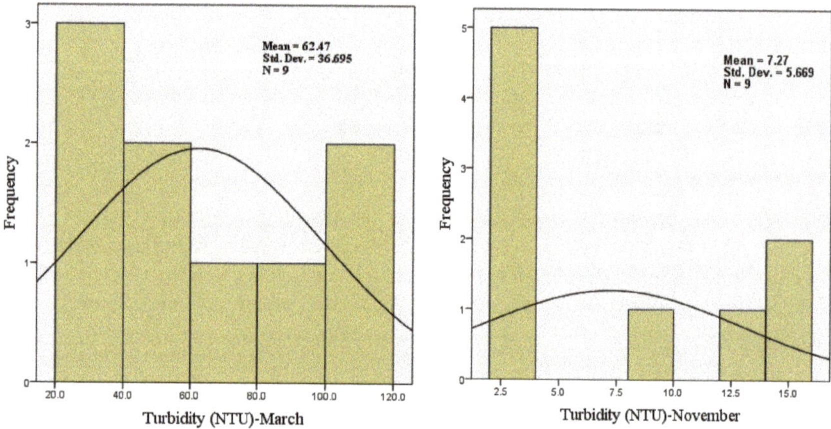

Figure 3. Frequency graphs showing the turbidity distribution on 13 March (**left**) and 27 November 2014 (**right**).

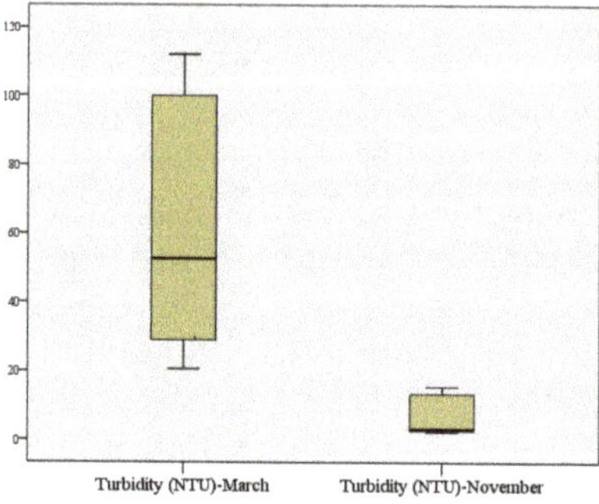

Figure 4. Box plots showing the spatial variability in turbidity on 13 March and 27 November 2014.

2.5. Image Acquisition

In this study, two L8SR images from 6 March and 17 November 2014 were used to retrieve turbidity in the Ramganga River. The selected images, with path 145 and row 40, were downloaded from the

United States Geological Survey (USGS) websites (http://earthexplorer.usgs.gov/). Each downloaded image was in a compressed folder containing TIFF images of each band.

2.6. Methodology

2.6.1. Rescaling

Rescaling of original L8SR bands wasapplied, as the range of the data wasfrom −2000–16,000. The valid range of reflectance is between 0–1. The minimum fraction of irradiance to be reflected from any surface should be 0.0 if it is a fully absorbed material, while the maximum fraction is 1.0 if it is a fully reflective material. The data were rescaled for the valid ranges, according to the information (Table 1) by multiplying each band by the scale factor of 0.001.

2.6.2. Masking

Only the river water body should be retained, and the rest needs to be masked. Masking of the water body was difficult, as the river is very narrow and it has many bridges. In addition, some areas of water in the river have been isolated in the form ofoxbow lakesthat appear after a broad meander from the main channel of the river is cut off, creating a free-standing body of water. Imagery masking was performed using version 10.2.2 of ArcGIS software. The river wasidentified by thresholding the images of the spectral reflectance.

2.7. Regression Models

The relationship between the L8SR reflectance and insitu measurements was developed by exploiting a simple linear backward elimination method. The backward elimination method begins with all the variables observed in the model. At each step, the least significant variable is removed. This process continues until there are no more insignificant variables. The user defines the level of significance at which the variables can be removed from the model [49]. In this study, IBM SPSS programming statistics v. 23.0 (Armonk, NY, USA), was used. Figure 5 shows the outline of the methodology applied in the present study. A regression model between the measured turbidity and the surface reflectance was applied. The output model has been validated, and the final results werethematic maps. For March, the regression was determined between the insitu turbidity on 13 March 2014 and the surface reflectance on 6 March 2014, while for November, the regression was between the insitu turbidity on 21 November 2014 and the surface reflectance on 17 November 2014.

Water quality indicators, such as turbidity, chlorophyll, and temperature, as well as suspended matter, have been retrieved from remote sensing, according to [22]. The following four types of expressions have been used to show the general forms of these experimental equations:

$$Y = A + BX$$

$$Y = ABX$$

$$Y = A + B \ln X$$

$$\ln Y = A + BX$$

where X is the measurement from remote sensing (i.e., radiance, reflectance, and energy); Y representswater quality parameters; A and B are empirically derived factors; and X could be energy, reflectance, orradiance in a single or two-band ratio. This concept has been adopted by many researchers in the past to retrieve the parameters of water quality; therefore in the present study, we followed the same concept, constructing an algorithm for turbidity retrieval that is dependent on the relationship between L8SR and insitu observation.

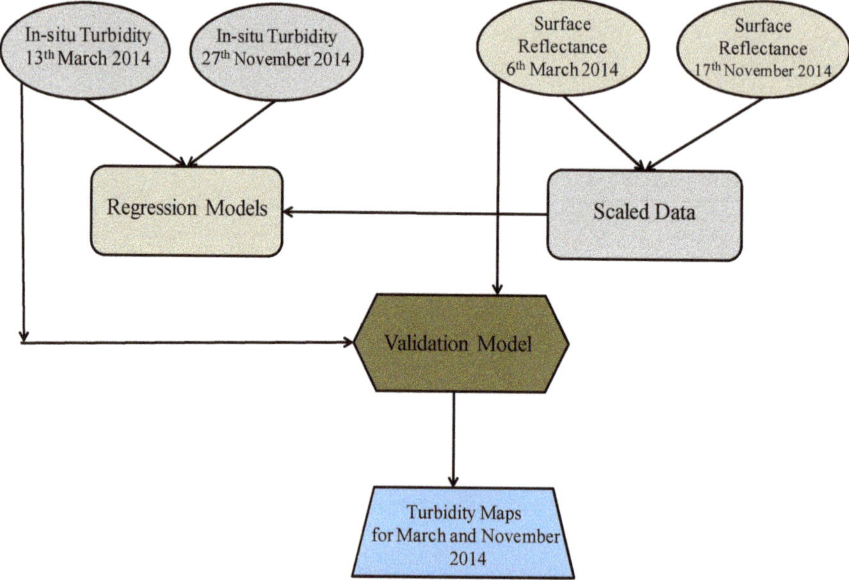

Figure 5. Flow chart summarizing the approaches followed.

3. Results

3.1. Retrieval of Turbidity

Statistical techniques for the derivation of chlorophyll- a (Chl-a) concentration and turbidity have been a common approach, based on the correlation between insitu data and spectral band values. The derived algorithms can provide an adequate estimate of Chl-a concentration [50] and turbidity [51]. These techniques were also adopted in the Ramganga River, in order to combine in situ data with satellite data to retrieve turbidity. The correlation was pursued between the insitu turbidity data and L8SR (single and ratio bands) for March and November 2014. After testing more than 20 band combinations in this correlation analysis, all single bands showed very poor correlation coefficients. Similar results appeared with different band ratios, except for b2/b3 and b2/b4, which produced higher coefficients of determination. The most significant results are presented in Table 4. Our results agree well with the findings of [52,53], who used b2/b3 and b2/b4 for the retrieval of turbidity from surface reflectance.

Using backward linear regression for the March data, all insignificant bands were removed, and the predictive model results were 0.75 and 0.56 for correlation coefficient (R) and R^2 values, respectively—whereas, for November, the value of R and R^2 were 0.852 and 0.726, respectively. However, the absence of autocorrelation in the residuals was indicated by Durbin–Watson's statistic (Tables S1 and S2). The description and summary of the final models of water quality parameters are shown in Table 5.

Table 4. Correlation between bands and reflectance for March and November 2014.

Bands	March	November
b2	0.051	−0.39
b3	−0.141	0.196
b4	−0.209	0.069
b5	−0.416	−0.155
logb2	0.045	−0.402
logb3	−0.153	0.187
b2b3	0.581	−0.852 **
b2b4	0.748 *	−0.756 *
b2b5	0.424	0.101
b3b4	0.523	0.391
b3b5	0.372	0.360
log(b3/b5)	0.389	0.280
b4/b3	−0.530	−0.38
b4/b5	0.348	0.331
b5b4	−0.363	−0.126
log(b5/b3)	−0.389	−0.28
log(b5/b4)	−0.360	−0.236
logb2	0.045	−0.402
logb3	−0.153	0.187
logb4	−0.211	0.045

** Correlation is significant at the 0.01 level. * Correlation is significant at the 0.05 level.

Table 5. Statistical summary and description of the final turbidity model that was computed for March and November 2014.

	Model	R	R^2	Std.Error of the Estimate	R^2 Change	Durbin–Waston
March 2014	−1.1 + 5.8 (b2/b4)	0.75	0.56	0.2	−0.08	1.36
November 2014	3.896 − 4.186 (b2/b3)	0.852	0.687	0.202	−0.002	1.972

3.2. Algorithm Validation

Comparisons between the measured and predicted turbidity for the nine samples that were used to determine the turbidity model are shown in Figures 6 and 7 and Table S3, along with squared residual and root mean square error (RMSE). Moderate correlation factors(R^2) of 0.56 and 0.726, with RMSE 1.013 and 0.178, wereobtained for March and November, respectively. For March, the predicted turbidity ranged from 2.329 to 3.023 NTU in relation to the measured turbidity, which ranged from 1.31 to 2.049 NTU (Figure 6). For November, the predicted turbidity varied from 0.337 to 1.33 NTU, compared with 0.362 to 1.18 NTU for the measured values (Figure 7).

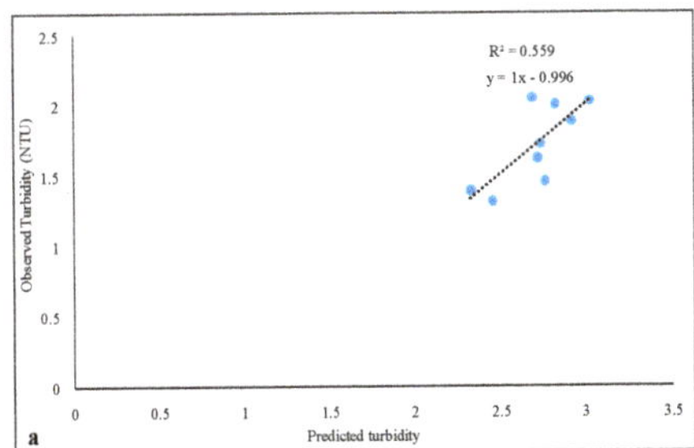

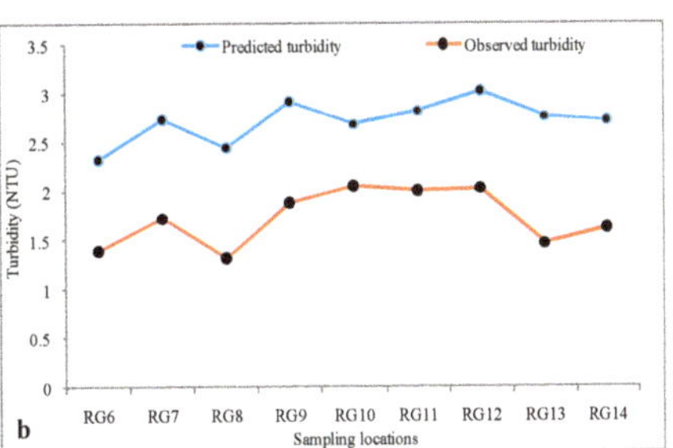

Figure 6. (**a**) Scatter plots and (**b**) line graphs show the comparison of observed and satellite-retrieved turbidity values at nine sampling sites from the Ramganga River in March 2014.

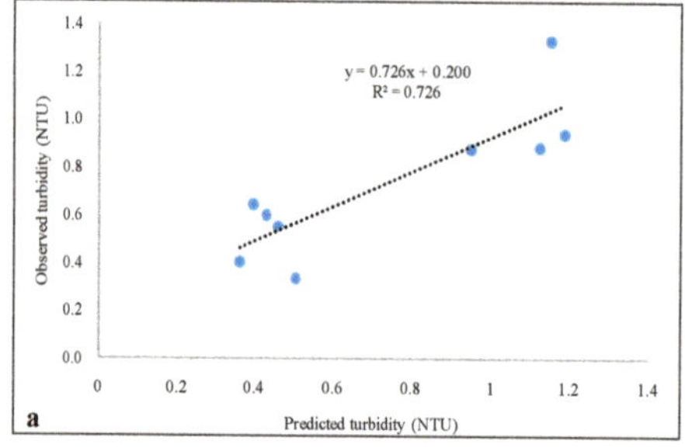 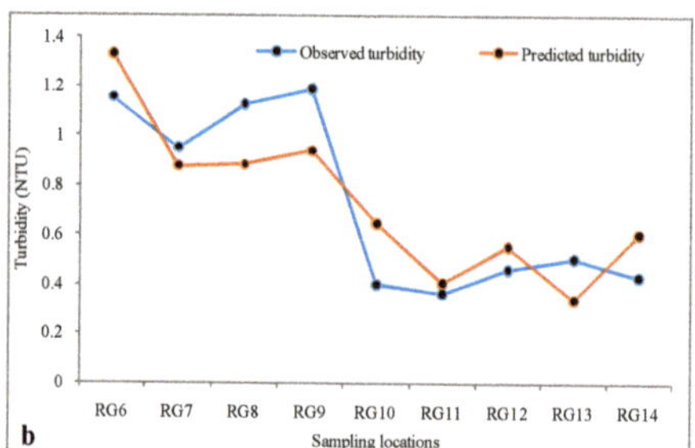

Figure 7. (**a**) Scatter plots and (**b**) line graphs shows the comparison of observed and satellite-retrieved turbidity values at nine sampling sites from the Ramganga River in November 2014.

3.3. Additional Validation for the Retrieved Model

For more precision inthe model, the four samples (RG2–RG5), which werenot included in the analysis, wereused to validate the model (Table 6). The first value in March and the last value in November weretoo high, because we tried to collect water samples where the water condition was rather uniform. However, it is still possible that the water samples capture locally high turbidity, while the reflectance of the satellite is on average about 900 m^2.

Table 6. Validation of the retrieved model in March and November 2014.

	Observed (NTU)	Predicted (NTU)	Square Residual	RMSE
March 2014	5.600	2.28803	10.969	2.2
	2.820	2.14487	0.456	
	0.888	1.45088	0.317	
	5.270	5.45901	0.036	
November 2014	1.2	2.204	1.008	1.39044
	0.5	0.854	0.125	
	0.6	1.331	0.535	
	3.5	1.3641	4.562	

The final turbidity maps, after applying the generated models, are presented in Figures 8 and 9, Figures S1a–c and S2a–c. For March, the estimated concentrations ranged from 2.329 to 3.023 NTU (Figure 8 and Figure S1a–c) in relation to the in situ concentration turbidity, which ranged from 1.31 to 2.049 NTU. For November, the estimated concentrations ranged from 0.337 to 1.33 NTU, compared to 0.362 to 1.18 NTU for the measured values (Figure 9 and Figure S2a–c).

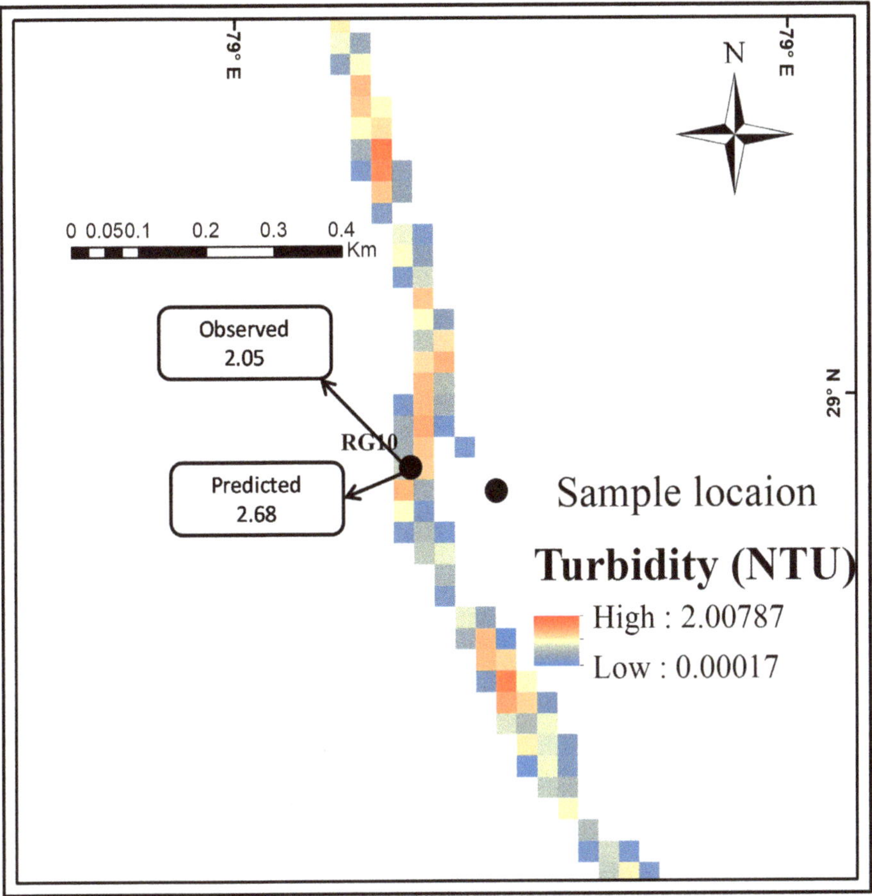

Figure 8. Turbidity map of the Ramganga River on 11 March 2014.

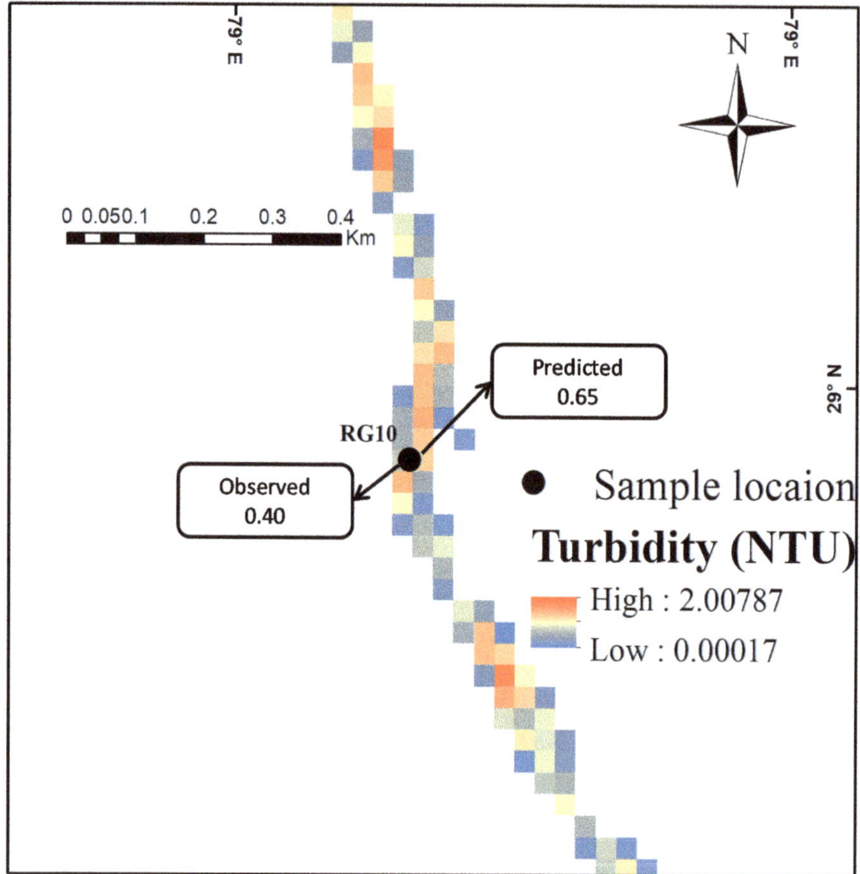

Figure 9. Turbidity map of the Ramganga River on 11 November 2014.

4. Discussion

The main objective of this study was to construct an algorithm to retrieve turbidity in the Ramganga River using L8SR. Statistical techniques [54,55] have been applied to determine the relationship between surface reflectance and measured turbidity. Bands from b1–b5 showed weak correlations for March and November. Different band ratios were utilized—for example, b2/b5, b3/b4, b3/b5, and b4/b5. The b2/b4 ratio was observed to be the most proficient for the estimation of turbidity for March, whereas b2/b3 was the most effective ratio for the estimation of turbidity in November for the Ramganga River. That was because vegetation indices (VIS) (b1, b2, and b3) and near infrared (NIR) (b5) are the most sensitive bands to SSC changes in water surface [56].

Such a monitoring system by remote sensing could be used as an early forewarning system for turbidity exceedance, which could help to make timely decisions about allowed emissions into the river water. Thus, simple and less expensive regular monitoring can be applied at a considerably larger spatial scale than continuous conventional sampling methods. However, errors related to satellite data, which reduce the accuracy of the resulting maps, are as follows:

- The samples collected may not be representative in relation to the total area of the water body;
- Water contains many soluble substances that hinder the process of obtaining the precise signature of the studied parameters;

- The difference in date between the acquisition of the satellite data and the insitu data;
- The relatively low spatial resolution of satellite images may affect their accuracy;
- The uncertainty of the locations of the pixels and insitu samples;
- The small number of samples affects the regression model, as well as the validation process.

A major problem with medium-resolution satellite data like Landsat 8 is that the Ramganga River is irregular in shape, generally narrow (about 100 m wide), and includes small islands. The reflected radiation from the shore and the vegetation near the shore is generally stronger than the radiation from the water. Therefore, the retrieval water quality parameters might not be possible if even a small portion of a pixel is covered with land. Also, the distinguished turbidity models are probably not relevant for different streams, and are along these site-specific lines. In all cases, testson freshwater bodies with comparable attributes should be undertaken to access the suitability of the models.

5. Conclusions

To retrieve surface turbidity from the L8SR product, a regional algorithm was developed and used in the Ramganga River. This investigation suggests that satellite information can be a ground-breaking device to foresee the concentration of turbidity in stream waters, and particularly in the Ramganga River. However, the distinguished models would be efficient only in the Ramganga River or rivers with comparable water quality and morphological characteristics. Nevertheless, even with the existence of a lot of ground information similar to the case in our examination, a quantitatively accurate estimation of water quality components in inland waters is a great challenge. Using the data acquired byvarious other sensors, such as Sentinel 2, Moderate Resolution Imaging Spectroradiometer (MODIS), and Gaofen-3 (GF-3),can help improve our ability to correctly estimate surface water characteristics from space.

Supplementary Materials: The following are available online at http://www.mdpi.com/2076-3417/10/11/3702/s1, Figure S1a: Turbidity map of Ramganga River in 6 March 2014, Figure S1b: Turbidity map of Ramganga River in 6 March 2014, Figure S1c: Turbidity map of Ramganga River in 6 March 2014, Figure S2a: Turbidity map of Ramganga River in 11 November 2014, Figure S2b: Turbidity map of Ramganga River in 11 November 2014, Figure S2c: Turbidity map of Ramganga River in 11 November 2014, Table S1: Models' summary and regression analysis statistics among turbidity concentrations and surface reflectance values for March and November 2014 (dependent variable), Table S2: Variables entered/removed from turbidity predictive models relying upon the regression method utilized for March and November 2014, Table S3: Comparison of satellites retrieved and in-situ observed turbidities values at 9 sampling sites of Ramganga River in March and November 2014 with statistical analysis for squared residual, root mean square (RMSE).

Author Contributions: Software: M.A. and M.Y.A.K.; Writing M.A. and M.Y.A.K.; revision and final editing: M.A., Q.M. and M.Y.A.K.; remote sensing and GIS data analysis: M.A. and Q.M.; In-situ data collection and analysis: M.Y.A.K. All authors have read and agreed to the published version of the manuscript.

Funding: This research was funded by the Council for Scientific and Industrial Research (CSIR), New Delhi, grant number 19-06/2011 (i) EU-IV.

Acknowledgments: The authors are also thankful to the National Key Research and Development Program, "Spatial information service and application demonstration of comprehensive monitoring of urban and rural ecological environment" (project number: 2017YFB0503900, 2017YFB0503905); the Major Projects of High Resolution Earth Observation Systems of National Science and Technology (05-Y30B01-9001-19/20-1); and the Science and Technology Cooperation Project of Sanya Municipal Institute, "Non-Point Source Pollution Risk Identification and Quantitative Assessment for Surface Water Source Using Remote sensing" (project number: 2018YD10), for their technical support in this study. The authors are also thankful to Council for Scientific and Industrial Research (CSIR), New Delhi, and IIT Roorkee, India for giving a research fellowship (ref. no: 19-06/2011 (i) EU-IV). The authors gratefully acknowledge the comments of the reviewers and the editor, which enormously improved the presentation of the manuscript.

Conflicts of Interest: The authors declare no conflict of interest.

References

1. Zheng, G.J. *Hydrodynamics and Water Quality: Modeling Rivers, Lakes, and Estuary*; John Wiley & Sons, Inc.: Hoboken, NJ, USA, 2007; pp. 130–134.

2. Güttler, F.N.; Niculescu, S.; Gohin, F. Turbidity retrieval and monitoring of Danube Delta waters using multisensor optical remote sensing data: An integrated view from the delta plain lakes to the western–northwestern Black Sea coastal zone. *Remote Sens. Environ.* **2013**, *132*, 86–101. [CrossRef]

3. Zablotskii, V.R.; Le, T.G.; Dinh, T.T.H.; Le, T.T.; Trinh, T.T.; Nguyen, T.T.N. Estimation of suspended sediment concentration using vnredsat–1A multispectral data, a case study in red river, hanoi, vietnam. *Geogr. Environ. Sustain.* **2018**, *11*, 49–60.

4. Chalov, S.; Golosov, V.; Tsyplenkov, A.; Theuring, P.; Zakerinejad, R.; Märker, M.; Samokhin, M. A toolbox for sediment budget research in small catchments. *Geogr. Environ. Sustain.* **2017**, *10*, 43–68. [CrossRef]

5. Chen, S.; Fang, L.; Zhang, L.; Huang, W. Remote sensing of turbidity in seawater intrusion reaches of Pearl River Estuary–A case study in Modaomen water way, China. *Estuar. Coast. Shelf Sci.* **2009**, *82*, 119–127. [CrossRef]

6. Minnesota Pollution Control Agency. Turbidity: Description, Impact on Water Quality, Sources, Measures—A General Overview, USA. 2008. Available online: https://www.pca.state.mn.us/sites/default/files/wq-iw3-21.pdf (accessed on 25 June 2019).

7. Kumar, A.; MMS, C.P.; Chaturvedi, A.K.; Shabnam, A.A.; Subrahmanyam, G.; Mondal, R.; Yadav, K.K. Lead Toxicity: Health Hazards, Influence on Food Chain, and Sustainable Remediation Approaches. *Int. J. Environ. Res. Public Health* **2020**, *17*, 2179. [CrossRef] [PubMed]

8. Mishra, S.; Kumar, A. Estimation of physicochemical characteristics and associated metal contamination risk in river Narmada, India. *Environ. Eng. Res.* **2020**, *26*, 190521. [CrossRef]

9. Kumar, A.; Sharma, M.P.; Yang, T. Estimation of carbon stock for greenhouse gas emissions from hydropower reservoirs. *Stoch. Environ. Res. Risk Assess.* **2018**, *32*, 3183–3193. [CrossRef]

10. Aksnes, D.L.; Dupont, N.; Staby, A.; Fiksen, Ø.; Kaartvedt, S.; Aure, J. Coastal water darkening and implications for mesopelagic regime shifts in Norwegian fjords. *Mar. Ecol. Prog. Ser.* **2009**, *387*, 39–49. [CrossRef]

11. Carson, A.B.; Benjamin, M.J.; Krista, K.B.; Daniel, B.Y.; Christian, E.Z. Reconstructing turbidity in a glacially influenced lake using the Landsat TM and ETM+ surface reflectance climate data record archive, Lake Clark, Alaska. *Remote Sens.* **2015**, *7*, 13692–13710.

12. Khan, M.Y.A.; Gani, K.M.; Chakrapani, G.J. Assessment of surface water quality and its spatial variation. A case study of Ramganga River, Ganga Basin, India. *Arab. J. Geosci.* **2016**, *9*, 28.

13. Khan, M.Y.A.; Khan, B.; Chakrapani, G.J. Assessment of spatial variations in water quality of Garra River at Shahjahanpur, Ganga Basin, India. *Arab. J. Geosci.* **2016**, *9*, 516. [CrossRef]

14. Khan, M.Y.A.; Gani, K.M.; Chakrapani, G.J. Spatial and temporal variations of physicochemical and heavy metal pollution in Ramganga River—A tributary of River Ganges, India. *Environ. Earth Sci.* **2017**, *76*, 231. [CrossRef]

15. Khan, M.Y.; Hu, H.; Tian, F.; Wen, J. Monitoring the spatio-temporal impact of small tributaries on the hydrochemical characteristics of Ramganga River, Ganges Basin, India. *Int. J. River Basin Manag.* **2019**, *18*, 231–241. [CrossRef]

16. Gernez, P.; Barille, L.; Lerouxel, A.; Mazeran, C.; Lucas, A.; Doxaran, D. Remote sensing of suspended particulate matter in turbid oyster farming ecosystems. *J. Geophys. Res. Oceans* **2014**, *119*, 7277–7294. [CrossRef]

17. Khan, M.Y.A.; Tian, F. Understanding the potential sources and environmental impacts of dissolved and suspended organic carbon in the diversified Ramganga River, Ganges Basin, India. *Proc. Int. Assoc. Hydrol. Sci.* **2018**, *379*, 61–66. [CrossRef]

18. Kumar, A.; Yang, T.; Sharma, M.P. Greenhouse gas measurement from Chinese freshwater bodies: A review. *J. Clean. Prod.* **2019**, *233*, 368–378. [CrossRef]

19. Kumar, A.; Sharma, M.P.; Taxak, A.K. Analysis of water environment changing trend in Bhagirathi tributary of Ganges in India. *Desal. Water Treat.* **2017**, *63*, 55–62. [CrossRef]

20. Kumar, A.; Sharmab, M.P.; Raic, S.P. A novel approach for river health assessment of Chambal using fuzzy modeling, India. *Desal. Water Treat.* **2017**, *58*, 72–79. [CrossRef]

21. Dogliotti, A.I.; Ruddick, K.G.; Nechad, B.; Doxaran, D.; Knaeps, E. A single algorithm to retrieve turbidity from remotely-sensed data in all coastal and estuarine waters. *Remote Sens. Environ.* **2015**, *156*, 157–168. [CrossRef]

22. Ritchie, C.J.; Zimba, V.P.; Everitt, H.J. Remote sensing techniques to assess water quality. *Photogramm. Eng. Remote Sens.* **2003**, *69*, 695–704. [CrossRef]

23. Allan, M.G.; Hamilton, D.P.; Hicks, B.J.; Brabyn, L. Landsat remote sensing of chlorophyll a concentrations in central North Island lakes of New Zealand. *Int. J. Remote Sens.* **2011**, *32*, 2037–2055. [CrossRef]

24. Tassan, S. A procedure to determine the particulate content of shallow water from Thematic Mapper data. *Int. J. Remote Sens.* **1998**, *19*, 557–562. [CrossRef]

25. Khan, M.Y.A.; Chakrapani, G.J. Particle size characteristics of Ramganga catchment area of Ganga River. In *Geostatistical and Geospatial Approaches for the Characterization of Natural Resources in the Environment*; Janardhana, R.N., Ed.; Springer: Cham, Switzerland, 2016; pp. 307–312.

26. Khan, M.Y.A. Spatial variation in the grain size characteristics of sediments in Ramganga River, Ganga Basin, India. In *Handbook of Environmental Materials Management*; Hussain, C.M., Ed.; Springer: Berlin, Germany, 2018.

27. Khan, M.Y.A.; Daityari, S.; Chakrapani, G.J. Factors responsible for temporal and spatial variations in water and sediment discharge in Ramganga River, Ganga Basin, India. *Environ. Earth Sci.* **2016**, *75*, 283. [CrossRef]

28. Vanhellemont, Q.; Ruddick, K. Turbid wakes associated with offshore wind turbines observed with Landsat 8. *J. Remote Sens. Environ.* **2014**, *145*, 105–115. [CrossRef]

29. Choodarathnakara, A.L.; Kumar, T.A.; Koliwad, S.; Patil, C.G. Mixed pixels: A challenge in remote sensing data classification for improving performance. *Int. J. Adv. Res. Comput. Eng. Technol. (IJARCET)* **2012**, *1*, 261.

30. Kishino, M.; Tanaka, A.; Joji, I. Retrieval of Chlorophyll a, suspended solids, and colored dissolved organic matter in Tokyo Bay using ASTER data. *Remote Sens. Environ.* **2005**, *99*, 66–74. [CrossRef]

31. Lim, H.S.; MatJafri, M.Z.; Abdullah, K.; Asadpour, R. A Two-Band algorithm for total suspended solid concentration mapping using theos data. *J. Coast. Res.* **2012**, *29*, 624–630.

32. Tebbs, E.J.; Remedios, J.J.; Harper, D.M. Remote sensing of chlorophyll-a as a measure of cyanobacterial biomass in Lake Bogoria, a hypertrophic, saline–alkaline, flamingo lake, using Landsat ETM+. *Remote Sens. Environ.* **2013**, *135*, 92–106. [CrossRef]

33. Doxaran, D.; Froidefond, J.M.; Castaing, P.; Babin, M. Dynamics of the turbidity maximum zone in a macrotidal estuary (the Gironde, France): Observations from field and MODIS satellite data. *Estuar. Coast. Shelf Sci.* **2009**, *81*, 321–332. [CrossRef]

34. Petus, C.; Chust, G.; Gohin, F.; Doxaran, D.; Froidefond, J.M.; Sagarminaga, Y. Estimating turbidity and total suspended matter in the Adour River plume (South Bay of Biscay) using MODIS 250-m imagery. *Cont. Shelf Res.* **2010**, *30*, 379–392. [CrossRef]

35. Ouillon, S.; Douillet, P.; Petrenko, A.; Neveux, J.; Dupouy, C.; Froidefond, J.M.; Andréfouët, S.; Caravaca, A.M. Optical algorithms at satellite wavelengths for Total Suspended Matter in tropical coastal waters. *Sensor* **2008**, *8*, 4165–4185. [CrossRef]

36. Tong, P.H.S.; Truong, M.C.; Hoang, C.T. Detecting chlorophyll-a concentration and bloom patterns at upwelling area in South central Vietnam by high resolution multi-satellite data. *J. Environ. Sci. Eng. A* **2015**, *4*, 215–224.

37. Ali, P.Y.; Jie, D.; Sravanthi, N. Remote sensing of chlorophyll-a as a measure of red tide in Tokyo Bay using hotspot analysis. *J. Remote Sens. Appl. Soc. Environ.* **2015**, *2*, 11–25.

38. Zhang, Y.; Zhang, Y.; Shi, K.; Zha, Y.; Zhou, Y.; Liu, M. A Landsat 8 OLI-Based, semianalytical model for estimating the total suspended matter concentration in the slightly turbid Xin'anjiang reservoir (China). *IEEE J. Sel. Top. Appl. Earth Obs. Remote Sens.* **2016**, *9*, 398–413. [CrossRef]

39. Daityari, S.; Khan, M.Y. Temporal and spatial variations in the engineering properties of the sediments in Ramganga River, Ganga Basin, India. *Arab. J. Geosci.* **2017**, *10*, 134. [CrossRef]

40. Khan, M.Y.A.; Hasan, F.; Panwar, S.; Chakrapani, G.J. Neural network model for discharge and water-level prediction for Ramganga River catchment of Ganga Basin, India. *Hydrol. Sci. J.* **2016**, *61*, 2084–2095. [CrossRef]

41. Khan, M.Y.A.; Hasan, F.; Tian, F. Estimation of suspended sediment load using three neural network algorithms in Ramganga River catchment of Ganga Basin, India. *Sustain. Water Resour. Manag.* **2019**, *5*, 1115–1131. [CrossRef]

42. Khan, M.Y.A.; Tian, F.; Hasan, F.; Chakrapani, G.J. Artificial neural network simulation for prediction of suspended sediment concentration in the River Ramganga, Ganges Basin, India. *Int. J. Sediment Res.* **2019**, *34*, 95–107. [CrossRef]

43. CWC. *Environmental Evaluation Study of Ramganga Major Irrigation Project*; Central Water Commission: Uttar Pradesh, India, 2012; Volume 1, p. 16.

44. Gupta, R.P.; Joshi, B.C. Landslide hazard zoning using the GIS approach—A case study from the Ramganga catchment, Himalayas. *Eng. Geol.* **1990**, *28*, 119–131. [CrossRef]

45. Khan, A.U.; Rawat, B.P. *Quaternary Geology and Geomorphology of a Part of Ganga Basin in Parts of Bareilly, Badaun, Shahjahanpur and Pilibhit District, Uttar Pradesh*; Geological Survey of India (GSI): Kolkata, India, 1992.

46. American Public Health Association (APHA). *Standard Methods for the Examination of Water and Wastewater*, 20th ed.; American Public Health Association: Washington, DC, USA, 1998.

47. Islam, M.R.; Yamaguchi, Y.; Ogawa, K. Suspended sediment in the Ganges and Brahmaputra Rivers in Bangladesh: Observation from TM and AVHRR data. *Hydrol. Process.* **2001**, *15*, 493–509. [CrossRef]

48. Kaliraj, S.; Chandrasekar, N.; Mages, N.S. Multispectral image analysis of suspended sediment concentration along the Southern coast of Kanyakumari, Tamil Nadu, India. *J. Coast. Sci.* **2014**, *1*, 63–71.

49. Borboudakis, G.; Tsamardinos, I. Forward-backward selection with early dropping. *J. Mach. Learn. Res.* **2019**, *20*, 276–314.

50. Mayo, M.; Gitelson, A.; Yacobi, Y.Z.; Ben-Avraham, Z. Chlorophyll distribution in lakeKinneret determined from Landsat thematic mapper data. *Int. J. Remote Sens.* **1995**, *16*, 175–182. [CrossRef]

51. Fraser, R.N. Hyperspectral remote sensing of turbidity and chlorophyll a among Nebraska Sand Hills lakes. *Int. J. Remote Sens.* **1998**, *19*, 1579–1589. [CrossRef]

52. Bande, P.; Adam, E.; Elbasit, M.A.A.; Adelabu, S. Comparing landsat 8 and sentinel-2 in mapping water quality at vaal dam. In Proceedings of the International Geoscience and Remote Sensing Symposium (IGARSS), Valencia, Spain, 22–27 July 2018.

53. Markogianni, V.; Dimitriou, E.; Tzortziou, M. Monitoring of chlorophyll-a and turbidity in Evros River (Greece) using Landsat imagery. In Proceedings of the First International Conference on Remote Sensing and Geoinformation of the Environment (RSCy2013) International Society for Optics and Photonics, Paphos, Cyprus, 8–10 April 2013; Volume 8795, p. 87950R.

54. Lathrop, R.G.; Lillesand, T.M. Use of thematic mapper data to assess water quality in Green Bay and central Lake Michigan. *Photogramm. Eng. Remote Sens.* **1986**, *52*, 671–680.

55. Baban, S.M.J. Detecting water quality parameters in the Norfolk Broads, U.K. using Landsat imagery. *Int. J. Remote Sens.* **1993**, *14*, 1247–1267. [CrossRef]

56. Yepez, S.; Laraque, A.; Martinez, J.M.; De Sa, J.; Carrera, J.M.; Castellanos, B.; Lopez, J.L. Retrieval of suspended sediment concentrations using Landsat-8 OLI satellite images in the Orinoco River (Venezuela). *Comptes Rendus Geosci.* **2018**, *350*, 20–30. [CrossRef]

Article

Enhanced Phosphorus Removal from Wastewater Using RSPRC and a Novel Reactor

Yan Liu, Limin Zhang * and Rajendra Prasad Singh

School of Civil Engineering, Southeast University, Nanjing 210000, China; liuyian@seu.edu.cn (Y.L.); rajupsc@seu.edu.cn (R.P.S.)
* Correspondence: 220181172@seu.edu.cn; Tel.: +86-151-9590-8761

Received: 21 April 2020; Accepted: 18 May 2020; Published: 24 May 2020

Abstract: Fly ash and steel slag both have a good adsorption performance and many researchers have mixed the two to make effective adsorbents. Based on previous knowledge, activated clay is added in this study. In order to deep dephosphorize wastewater, two different industrial wastes (steel slag, fly ash) are blended into activated clay as adsorption substrates, supplemented with a binder and foaming agent to prepare a Residue and Soil Phosphorus Removal Composite (RSPRC). This is prepared to carry out experimental research on the decolorization effect and phosphorus removal characteristics of RSPRC. Meanwhile, a self-developed concentric circular diversion wall adsorption reactor is implemented to study the effect of phosphorus removal. It is found that the addition of activated clay can significantly improve the phosphorus removal performance. The results suggest that the phosphorus concentration in the effluent from the reactor can be stably reduced to below 0.10 mg/L. The concentric circular diversion wall adsorption reactor and RSPRC will have broad application prospects in phosphorus removal.

Keywords: phosphorus; adsorption; steel slag; fly ash; activated clay; reactor

---

1. Introduction

Phosphorus (P) usually originates from human and animal wastes, household detergents, food-processing effluents, commercial fertilizers, and agricultural land runoffs. Phosphorus is a major nutrient for biomass growth [1–3]. However, excessive concentrations of P in water bodies such as lakes, lagoons or rivers cause an abnormal growth of algae and aquatic plants resulting in the degradation of the water quality [4,5]. The State Environmental Protection Administration of China has recommended that total P should not exceed 0.5 mg/L according to the class A demands of discharge standard of pollutants for municipal wastewater treatment plants (GB18918-2002). The European Union (EU) maintains that the cut-off for total P concentration between risk and no risk of eutrophication in lakes is <10 µg/L to >100 µg/L, and for rivers, while total P concentrations below 0.01–0.07 µg/L are considered ideal. Therefore, the enhanced removal of phosphorus is a current trend. At present, removal methods of phosphorus from wastewater can be roughly divided into chemical precipitation [6], advanced biological treatment [7,8] and adsorption [9,10]. The biological phosphorus removal process is complicated and easily affected by the environment, which means that the effluent phosphorus concentration stability is poor [11]. Although the chemical removal method has a good effluent effect, most phosphorus removal agents used are industrial products such as lime, aluminum salt, iron salt, ferrous salt and magnesium salt, which lead to high phosphorus removal costs and a large amount of sludge. Thus, chemical precipitation has not been widely applied in practice [12]. Compared with other kinds of phosphorus removal methods, adsorption has the advantages of easy operation, simple process, reliable operation, high efficiency, low consumption, etc. [13]. Therefore, it has broad application prospects.

Studies over the past two decades have provided significant information on the profound phosphorus removal performance of steel slag and fly ash. However, there is clearly a lack of research related to the phosphorus removal performance of activated clay. Although extensive research has been carried out on steel slag and fly ash, there has been little research on combining active clay, steel slag and fly ash to remove phosphorus. Most studies in the field of phosphorus adsorption have only focused on the adsorbents. So far, very little attention has been paid to the role of reactors used for adsorption reactions. Currently, high efficiency phosphorus removal material with strong adsorption capacity, certain strength and chemical stability, low water flow resistance and low cost should be developed urgently and applied to the phosphorus removal process of various wastewaters to ensure the phosphorus removal effect on the wastewater and recover the phosphorus [14]. In addition, the development of a reactor that is easy to operate, has a large capacity for water treatment and can guarantee an excellent adsorption effect for a long period of time is also one of the hot topics in recent years.

Therefore, the prepared RSPRC was used as the adsorbent and a concentric circular diversion wall adsorption reactor was developed as the adsorption device. The main objectives of the current work are as follows: (1) assess the decolorization and dephosphorization performance of activated clay at different ratios to determine the optimum ratio of raw materials for preparing RSPRC. (2) Investigate the phosphorus removal characteristics of RSPRC. (3) Investigate the effect of RSPRC dosage on the phosphorus removal efficiency of the reactor. (4) Investigate the effect of RSPRC distribution on phosphorus removal efficiency to optimize the placement of RSPRC for adsorption reactors. Moreover, (5) investigate the effect of hydraulic retention time (HRT) in the reactor on the phosphorus removal efficiency.

2. Materials and Methods

2.1. The Concentric Circular Diversion Wall Adsorption Reactor

The design scheme of the concentric circular diversion wall adsorption reactor was as follows: comparing the advantages and disadvantages of the existing reactors, combined with the purpose of this research, the preliminary design scheme of the reactor type, namely the fixed bed type, was proposed. In order to realize the functions of multi-stage series adsorption and flexible distribution of adsorbents, taking the oxidation ditch as a reference, the reactor adopted the shape of concentric circles and was equipped with slots and grills. Based on the preliminary design scheme, comprehensively considering the advantages and disadvantages of various types of materials, after repeated research and discussion with the manufacturer on the details of the reactor, stainless steel was chosen as the reactor material and the necessary adjustments of the reactor structure were made. Finally, the reactor dimensions were determined, such as the height-to-diameter ratio of the reactor, the size of the water inlet and outlet, the size of the overflow port and so on.

The concentric circular diversion wall adsorption reactor was composed of an outlet pipe, an inlet pipe, an overflow port, a slot, removable grilles, diversion walls, diversion tubes, a constant flow pump and ball valves (Figure 1a,b). The total height of the reactor was 80 cm, the effective height was 50 cm and the support foot height was 30 cm. The diameter was 120 cm and the effective diameter was 100 cm. The inflow height was 78 cm and the effluent height was 28 cm. The total volume was about 400 L and the effective volume was approximately 375 L. The diameters of the diversion walls from the outside to the inside were 50 cm, 40 cm, 30 cm, 20 cm and 10 cm, respectively. The total heights of the diversion walls were 50 cm and the effective height was 48 cm, 46 cm, 44 cm, 42 cm and 40 cm from the outside to inside. The distances from the outside to the inside of the diversion wall were 2 cm, 4 cm, 6 cm, 8 cm and 10 cm. The dimensions of the removable grilles were: length × width × thickness = 50 cm × 10 cm × 1 cm; and the measured mesh area was about 0.04 cm^2. The outer and inner diameters of the diversion tube used in current work were 20 mm and 19 mm respectively.

(a)   (b)

Figure 1. (**a**) The concentric circular diversion wall adsorption reactor schematic; (**b**) the concentric circular diversion wall adsorption reactor.

2.2. RSPRC Preparation

The raw materials of RSPRC were divided into two types: substrate and auxiliary material. The substrate consisted of steel slag, fly ash and activated clay. The auxiliary material was cement as binder and the foaming agent to increase the specific surface area [15]. The steel slag and fly ash used in this experiment were taken from a steel plant and a power plant in Nanjing. The chemical composition of the steel slag and fly ash was analyzed by wavelength dispersive X-ray fluorescence analyzer (XRF-1800) produced by Shimadzu, Kyoto, Japan. Table 1 shows the chemical composition and content of the waste residue substrate and Table 2 presents the particle size distribution of the waste residue substrate.

Table 1. Percentage of oxides in waste residue substrate.

Oxide	CaO	Fe_2O_3	Al_2O_3	SiO_2	MgO	MnO	SO_3	V_2O_5	TiO_2	Na_2O	ZnO	CuO
Composition of fly ash (%)	1.31	4.39	45.9	44.4	0.261	0.026	0.666	0.038	1.26	0.094	0.021	0.02
Composition of steel slag (%)	55.0	21.5	1.51	13.4	3.65	1.75	0.512	0.417	0.296	0.077	-	-

Table 2. Size distribution of the waste residue substrate.

Particle Size of fly ash/mesh	<6	6~18	18~45	45~70	>70
proportion/%	7.45	12.76	42.93	23.55	11.87
Particle size of steel slag/mesh	<12	12~80	80~320	>320	-
proportion/%	2.49	7.68	32.40	57.36	-

The preparation of activated clay is divided into original soil treatment and modification treatment [16]. This experiment adopted acid modification treatment [17]. The bentonite was taken from Tangshan, Jiangning District, Nanjing, China. For the acid modification treatment, the bentonite was immersed in a 3 mol/L sulfuric acid solution, activated for 30 min, and stirred at a speed of 1000 r/min in a constant temperature environment of 100 °C. After the activation was completed, it was washed with distilled water to neutrality, dried, ground, and passed through a 100-mesh sieve to obtain experimental activated clay. Respectively, the principal oxides SiO_2, Al_2O_3, MgO, Fe_2O_3, and CaO in the activated clay account for 64.23%, 16.88%, 0.76%, 2.05% and 2.45%. The preparation of the adsorbent was carried out according to the different proportion of various substrates. The specific percentages of substrate and binder are presented in Table 3.

Table 3. Percentage composition of substrate and binder in Residue and Soil Phosphorus Removal Composite (RSPRC) in four groups.

Material	Fly Ash	Steel Slag	Activated Clay	Binder
Composition of Group 1 (%)	78.25	11.25	5	5
Composition of Group 2 (%)	74.38	10.63	10	5
Composition of Group 3 (%)	70	10	15	5
Composition of Group 4 (%)	83.13	11.88	—	5

A schematic flow chart of the RSPRC preparation process is provided in Figure 2. At first, the selected substrate (fly ash and steel slag) was mechanically pulverized. After that, an appropriate amount of activated clay, binder and foaming agent were blended and thoroughly mixed. Later, an appropriate amount of tap water was added to the materials and these materials were stirred evenly. Then, all the materials were poured into the drum granulator. After the granulation was completed, the water was sprinkled after 6~8 h. Finally, RSPRC was maintained for several days under normal temperature and pressure to keep certain humidity on the surface of the particles.

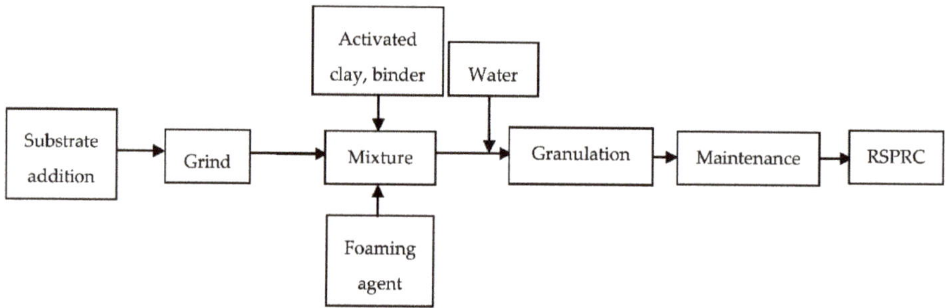

Figure 2. RSPRC preparation process.

According to the experimental conclusion of Wu et al. [18], the composite adsorbent with a particle size of 5 mm had a more efficient phosphorus removal effect than that with particle sizes of 10 and 20 mm. Therefore, the particle size selected for this experiment was 5 mm. Zhou [19] pointed out in their research on Efficient Phosphorus Removal Composite (EPRC)that the use of a plant-type foaming agent in the preparation process made it more difficult to cause secondary pollution, and that the optimal dosage was 8 mL/kg. RSPRC and EPRC are basically the same in terms of material physical properties. Consequently, the preparation of the adsorbent was carried out with an 8 mL/kg foaming agent. The plant-type foaming agent was bought from Shanghai Fangbao Building Material Technology Co., Ltd., Shanghai, China. The plant-type foaming agent is prepared from rosin and sodium hydroxide through a saponification reaction and had the advantages of its low cost and large foam production.

2.3. Study on Decolorization and P Removal of RSPRC with Activated Clay

The colored wastewater used in the experiment was derived from a river in the Jiulonghu Campus of Southeast University. A total of 250 mL wastewater was taken and stored in conical flasks. As shown in Table 3, 5 g adsorbents of different groups were respectively added. The colored wastewater was slowly stirred for 20 min and stood for 2 h. After that, the wastewater was filtered and the filtrate was taken to measure absorbance. The absorbance was measured by a spectrophotometer at a wavelength of 750 nm. The decolorization rate was calculated by the following formula:

$$Q = \frac{(A_0 - A_e)}{A_0} \times 100\% \tag{1}$$

where Q is the decolorization rate, A_0 is the initial absorbance of the colored wastewater and A_e is the absorbance of the colored wastewater after adsorption.

A total of 10 g of different groups of adsorbents were placed into 250 mL Erlenmeyer flasks. Then, the simulated solution, with a phosphorus concentration of 2 mg/L, prepared with deionized water and KH_2PO_4, was separately added to the Erlenmeyer flasks up to the mark. The pH of the simulated solution was adjusted to 7 by adding HCL and NaOH solution. Finally, they were oscillated in a constant temperature oscillator. When the adsorption duration was 60 min, 120 min and 180 min, the supernatant was taken, and the P concentrations were determined using the molybdenum blue spectrophotometric method stated in American Public Health Association (APHA) [20] in order to investigate the effect of the ratio of activated clay in the adsorbent on phosphorus removal.

The phosphorus removal effect of RSPRC in the advanced treatment of wastewater was mainly reflected in the phosphorus removal rate and the amount of phosphorus adsorbed by RSPRC. The adsorption capacity (Qe, mg/g) or amount of phosphorus adsorbed by RSPRC and removal rate (R) of phosphorus were calculated from the following equations:

$$Q_e = \frac{(C_0 - C_e)V}{m} \tag{2}$$

$$R\ (\%) = 100 \times \frac{C_0 - C_e}{C_0} \tag{3}$$

where Co is the initial concentration of the P (mg/L), Ce is the equilibrium or residual P concentration (mg/L), V is the volume of the solution (L) and m is the mass of adsorbent (g).

2.4. Study on P Removal Characteristics of RSPRC

At first, for the purpose of exploring the effect of dosage on the phosphorus removal of RSPRC, 2, 3, 4, 5, and 6 g RSPRC samples with a particle size of 5 mm were placed in five conical flasks, which contained 100 mL of KH_2PO_4 standard solution with a P concentration of 0.5 mg/L and pH = 7. The five conical flasks were shaken well in a 150 r/min shaker. After 24 h, the supernatant was taken and the P concentration was measured. Secondly, in order to study the effect of contact time, when the static adsorption times were 4, 8, 24, 72 and 144 h, the supernatant was taken and the P concentration was determined. Thirdly, 5 g of RSPRC was separately put into 1 L of simulated wastewater with pH = 7 and P concentrations of 0.3, 0.5 and 1.0 mg/L. The P concentration was detected at different times to investigate the effect of initial P concentration on the phosphorus removal of RSPRC.

2.5. Study on the Influence of RSPRC Distribution in the Reactor on Phosphorus Removal Effect

In an attempt to study the influence of different distributions of RSPRC in the reactor on the phosphorus removal effect, four different distributions were designed and a certain amount of adsorbent was placed in the reactor in the specified way. Four distributions of RSPRC are presented in Figure 3. The initial P concentration of the experimental influent was set to 0.5 mg/L, the pH was adjusted to 7, and the total amount of adsorbent was 10 kg, so as to better observe the phosphorus removal effect of the reactor under different distributions. Additionally, the hydraulic retention time was 3 h and the water flow rate was 125 L/h.

2.6. Study on the Influence of RSPRC Dosage in the Reactor on Phosphorus Removal Effect

In this experiment, the initial phosphorus concentrations of the wastewater were set to three gradients: 0.3, 0.4 and 0.5 mg/L. The pH of wastewater was adjusted to 7. At present, the reaction time required for the wastewater advanced treatment process is 1~6 h, and hence the adsorption reaction time set in this experiment was 2~4 h, which is far less than the previously set reaction balance time. Therefore, for the purpose of reaching the set target effluent P concentration (<0.1 mg/L), the experiment needed to increase the dosage of the adsorbent and shorten the time required for the static adsorption to reach equilibrium. The initial doses of the adsorbent for the entire reactor were 4, 6, 8, and 10 kg, which were numbered as group 1, group 2, group 3, and group 4. The distribution of the adsorbents was set between the diversion walls at equal intervals, as shown in experimental condition d in Figure 3, and the hydraulic retention time was set to 3 h. Finally, the effluent P concentration of each group was determined.

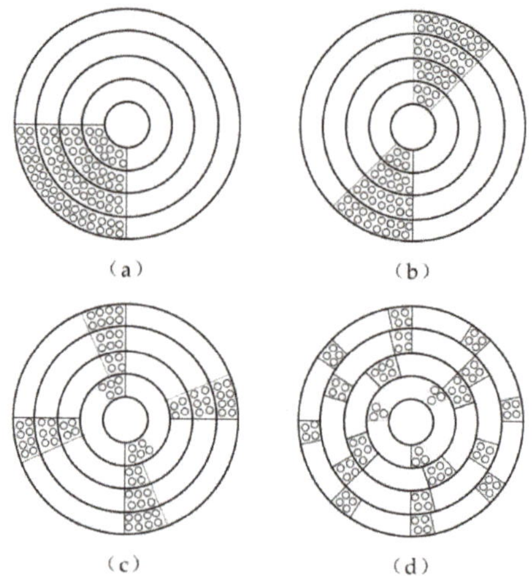

Figure 3. Adsorbent distribution of four experimental conditions (**a–d**).

2.7. Study on the Influence of HRT in the Reactor on Phosphorus Removal Effect

Keeping in mind the time required for the wastewater treatment process in the self-developed reactor, the initial phosphorus concentration of the statically adsorbed wastewater was set to 0.5 mg/L, and the hydraulic retention times were set to 2, 3 and 4 h, respectively. The pH of the wastewater was adjusted to 7. The P concentration in the effluent of each group was detected at the corresponding moment.

3. Results and Discussion

3.1. Effect of Proportion of Activated Clay in RSPRC on Decolorization and Phosphorus Removal

The experimental results reveal that the decolorization rate was only 46.7% without the addition of activated clay in RSPRC; following the addition of activated clay, a significant increase in the decolorization rate of the wastewater was recorded. When the proportion of activated clay was 15%, the decolorization rate was as high as 70.8%. This indicates that the adsorbent RSPRC is equipped with a better decolorization performance because of the addition of activated clay. The experimental results of exploring the phosphorus removal performance of activated clay are given in Figure 4. The phosphorus removal rate was 75.20% when the proportion of activated clay in group 3 was 15%. On the other hand, the phosphorus removal rate of the adsorbent containing no activated clay in group 4 was merely 68.74% at 180 min. The results show that there is a clear trend of an increasing phosphorus removal rate with the addition of activated clay. For instance, the phosphorus removal rate of the adsorbent with 15% increased by 6.46%. It is apparent that the addition of activated clay can effectively improve the phosphorus removal performance of the adsorbent. This result may be explained by the fact that the activated clay was obtained by rinsing and drying bentonite after it was activated by inorganic acid, and that its main composition is montmorillonite, which has a high specific surface area [21]. During the acid activation treatment, Wang et al. found that Na^+, Mg^{2+}, K^+, Ca^{2+} and other cations between the bentonite layers can be converted into soluble salts and dissolved out, increasing the layer spacing, and forming a porous active substance with a microporous mesh structure and a large specific surface area. Additionally, the impurities distributed in the bentonite structure channels can also be removed. The pore volume was increased, which is beneficial to the diffusion of adsorbate molecules [17].

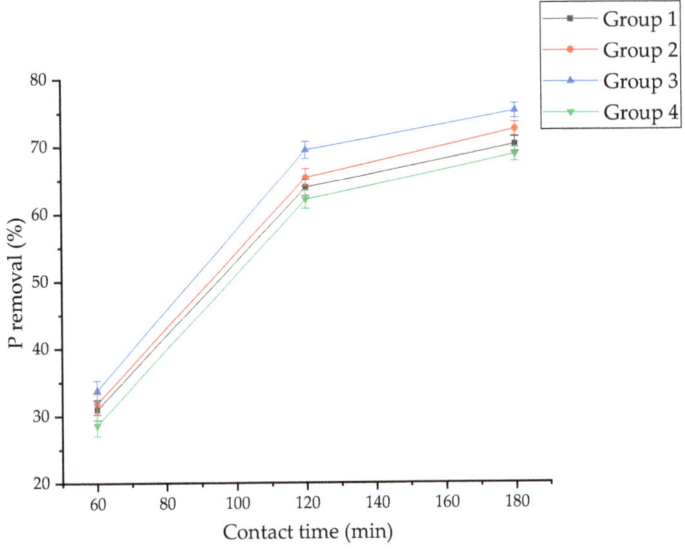

Figure 4. Effect of proportion of activated clay on Phosphorus removal efficiency.

Although the phosphorus removal performance of the adsorbent was improved with the increase in the proportion of activated clay in the substrate, the improvement was not obvious enough. Taking into account economic factors, the basic ratio of adsorbent materials was fly ash:steel slag:activated clay:cement = 14:2:3:1 in the subsequent experiments.

3.2. Characterization of RSPRC

The microstructure of RSPRC was observed by a Nova scanning electron microscope (SEM) using Nano SEM 450. Table 4 presents the physical properties of RSPRC used in experiments.

Table 4. Physical properties of RSPRC used in experiments.

Physical Parameter	Grain Size (mm)	Density (g/cm^3)	Porosity (%)	Specific Surface Area (m^2/g)
RSPRC	5~20	1.16	30.98	2.62~8.56

An SEM image of RSPRC is shown in Figure 5. The SEM image shows that RSPRC is composed of particles with different properties, of which spherical particles account for more than 60% (Figure 5a). After analysis, we established that these are the glass bodies in the fly ash, which have stored a high chemical internal energy after high-temperature calcination and are the source of the fly ash activity. Figure 5b shows that the glass body is a hollow sphere. When the sphere breaks, the Al_2O_3 and SiO_2 inside are released, the broken bond increases, the specific surface area increases and the reaction contact area increases, which increases the number of activated molecules and effectively improves the early chemical activity of the RSPRC. Figure 5c presents the fracture surface of the glass body. RSPRC contains a lot of minerals. Steel slag and activated clay are its main sources. They are $2CaO·SiO_2$, $3CaO·SiO_2$, $2FeO·SiO_2$, $2CaO·Fe_2O_3$ and free calcium oxide (f-CaO), etc., many of which possess a phosphorus removal ability (Figure 5d) [22].

Figure 5. SEM images of RSPRC. The magnifications from (**a–d**) are 5080, 2230, 4070, 26260.

3.3. Phosphorus Removal Characteristics of RSPRC

As can be seen from Figure 6, there has been a gradual rise in the phosphorus removal rate with the increasing adsorbent dosage. The phosphorus removal rate of RSPRC reached a peak when the adsorbent dosage was 6 g. However, the adsorption capacity (Qe, mg/g) or amount of phosphorus adsorbed by RSPRC had a downward trend when the dosage of adsorbent increased. Similarly, Ashekuzzaman et al. (2014) discovered that the phosphate sorption capacity was decreased with increasing dose of layered double hydroxides, because with the increasing dose, the adsorbent mass increased in the same volume of adsorbate solution, while the mass of the adsorbate to be sorbed remained the same [22]. As the dosage increased, although the available active sites were increasing, the increase in the density of the adsorbent caused the adsorption active sites to be stacked, so the adsorption efficiency was lowered, resulting in a slow increase in the removal rate of phosphorus and the decline in the amount of phosphorus adsorbed by RSPRC [23,24].

Figure 7 provides the results obtained from the experiment of studying the effect of contact time on the phosphorus removal of RSPRC. As the contact time increases, the effluent P concentration decreases and the phosphorus removal rate increases. With the increase in contact time, different dosages of adsorbent have similar changes in phosphorus removal rate. In the interval of 0~8 h, the effluent P concentration decreased slowly. In the case of the 8~24 h interval, there was a sharp drop in the effluent P concentration. However, in the 24~72 h interval, the effluent P concentration declined gradually, and after 72 h, the effluent P concentration tended to be stable and basically reached the adsorption equilibrium. The observed correlation between contact time and phosphorus removal effect might be explained as follows: there are more active sites on the surface and pores of the RSPRC, where the initial adsorption exists for a short time, so the phosphate ions rapidly occupy the active site, and the adsorption rate is faster. With the extension of time, RSPRC adsorbs more and more phosphorus from the wastewater and the active sites are fewer, which can result in a decline in the removal rate and the effluent P concentration [25]. Shan et al. (2009) also found that an almost 70% removal of phosphate by fly ash was reached in a short time and progressively increased with the contact time. However, there were no further increases in the percentage of phosphate removal after 20 h [26].

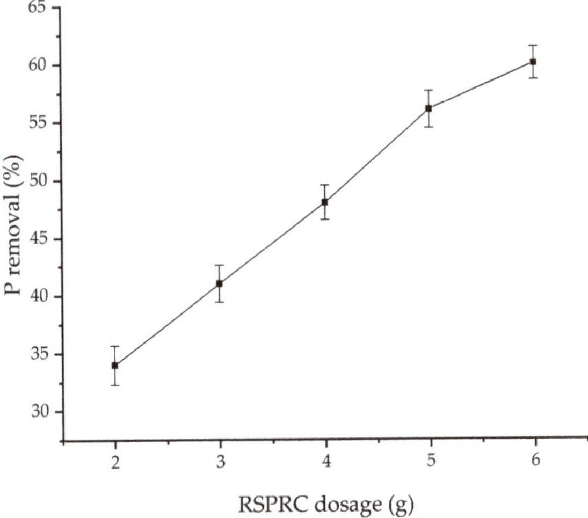

Figure 6. Effect of RSPRC dosage on phosphorus removal efficiency.

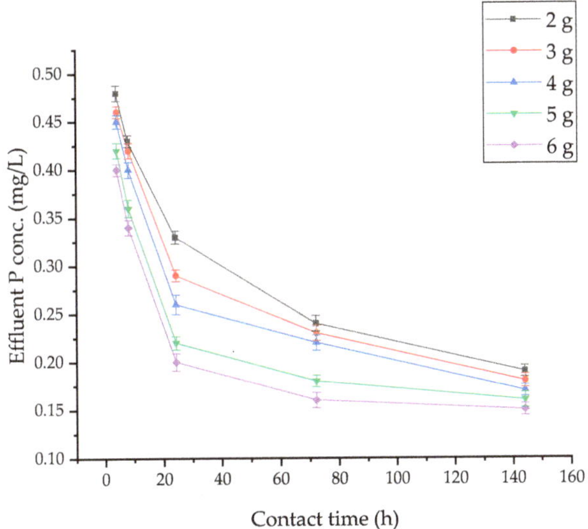

Figure 7. Effect of contact time on phosphorus removal efficiency.

The experimental results on the effect of initial phosphorus concentration on the phosphorus removal of RSPRC are presented in Figure 8. The adsorption capacity (Qe, mg/g) or amount of phosphorus adsorbed by RSPRC significantly increased when the initial P concentration increased. As the contact time is increased, the higher the initial P concentration is, the faster the amount of phosphorus adsorbed by RSPRC is increased. This result indicates that, under the reaction conditions with a high P concentration, the amount of phosphorus adsorbed by RSPRC increased and the efficiency of phosphorus removal was significantly improved [27]. These relationships may partly be explained by the fact that there are sufficient adsorption sites on the surface of the adsorbent. At the same time, the initial phosphate concentration in the solution is high, so a sufficient driving force for mass transport promotes the adsorption of phosphate by the RSPRC [28,29]. Meyer et al. obtained the result that the adsorption capacity of electric arc furnace steel slags ranged from 0.09 to 0.28 mg/g and did not increase according to initial P concentrations above 10 mg /L, suggesting that a limit in P removal was reached [30]. On the contrary, when the initial P concentration was from 0.3 to 1 mg/L, the adsorption capacity has been increasing, indicating that the initial P concentration in this experiment was not high enough to make the adsorption capacity reach the limit and that the driving force had not yet achieved its maximum.

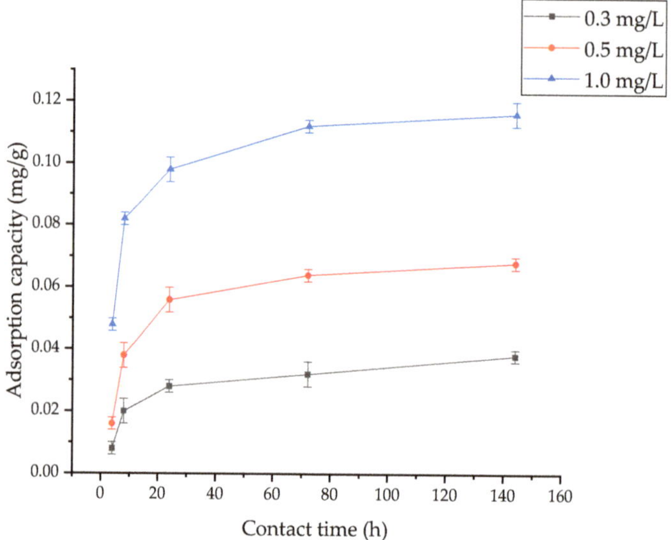

Figure 8. Effect of initial concentrations on the absorption of phosphorus.

3.4. RSPRC Adsorption Kinetics

When the adsorbent adsorbs the wastewater, its adsorption function is mainly realized by the chemical bond force, electrostatic attraction and Van der Waals force between the adsorbent surface molecules and the adsorbate molecules in the wastewater. Currently, pseudo-first order kinetic models and pseudo-second order kinetic models are often used to describe the dynamic behavior of solid–liquid static adsorption.

The pseudo-first order kinetic equation can express the relationship between the rate of adsorption reaction and the environmental conditions (such as concentration) involved in the adsorption reaction or related to the adsorption reaction. This kinetic model is usually based on the solid adsorption equilibrium capacity. The differential is as follows:

$$\frac{d_{qt}}{dt} = k_1\left(q_e - q_t\right) \tag{4}$$

After integrating the differential equation, the pseudo-first order kinetic equation can be expressed as follows:

$$\ln\left(q_e - q_t\right) = \ln q_e - k_1 t \tag{5}$$

where q_e is the adsorption amount of the adsorbent surface to the solute under adsorption equilibrium (mg/g), q_t is the adsorption amount of the adsorbent surface to the solute at the specified time (t) during the adsorption process (mg/g), and k_1 is the pseudo-first order kinetic equation adsorption rate constant.

The pseudo-second order kinetic model is applicable to the solid–liquid static adsorption reaction in the presence of a saturated site, which can effectively represent the composite effect under the action of multiple adsorption mechanisms. Generally, the kinetic model is based on the adsorption equilibrium capacity. The differential is as follows:

$$\frac{dq_t}{dt} = k_2\left(q_e - q_t\right) \tag{6}$$

After integrating the differential equation, the pseudo-second order kinetic equation can be expressed as follows:

$$\frac{t}{q_t} = \frac{1}{k_2 q_e^2} + \frac{1}{q_e}t \tag{7}$$

where q_e is the adsorption amount of the adsorbent surface to the solute under adsorption equilibrium (mg/g), q_t is the adsorption amount of the adsorbent surface to the solute at the specified time (t) during the adsorption process (mg/g), and k_2 is the pseudo-second order kinetic equation adsorption rate constant.

In general, there is a slight deviation when the single kinetic equation is used to reflect the actual reaction. Therefore, it is necessary to conduct a comparative study after fitting the pseudo-first order kinetic equation and the pseudo-second order kinetic equation. The relevant parameters of the pseudo-first order and pseudo-second order kinetic equations are presented in Table 5.

Table 5. Kinetics of phosphorus removal by RSPRC at different initial concentrations.

Initial Concentration (mg·L⁻¹)	q_{exp} (mg·g⁻¹)	Pseudo-First Order Kinetic Model			Pseudo-Second Order Kinetic Model		
		q_e (mg·g⁻¹)	K_1 (h⁻¹)	R^2	q_e (mg·g⁻¹)	K_2 (h⁻¹)	R^2
0.3	0.038	0.036	0.016	0.9246	0.039	0.301	0.9813
0.5	0.068	0.064	0.012	0.9638	0.072	0.416	0.9842
1.0	0.116	0.110	0.72	0.9644	0.118	0.742	0.9954

The pseudo-first order kinetic model and the pseudo-second order kinetic model were utilized to analyze the adsorption kinetics of the adsorbent at different initial phosphorus concentrations. The results illustrate that the correlation coefficients of the pseudo-first order kinetic equation and the pseudo-second order kinetic equation are more than 0.9 and can better reflect the phosphorus adsorption process in different initial P concentrations. The fitting results of the pseudo-second order kinetic model are more accurate than the pseudo-first order kinetic model, because the correlation coefficients exceeded 0.98, while the correlation coefficient of the pseudo-first order kinetic model was only 0.9246 at an initial P concentration of 0.3 mg/L. The adsorption amount of the adsorbent surface to the solute under the adsorption equilibrium (q_e) obtained by the pseudo-second order kinetic model was larger than that of the sample measured result (q_{exp}). The reason for the difference is that when the adsorption process continues, the P concentration of the wastewater continues to decrease, the conditions required for each adsorption mechanism change, the molecular force between the surface of the adsorbent and phosphorus is reduced, and the concentration difference required for the chemical adsorption decreases. These all lead to a decrease in the adsorption capacity.

In the second-order kinetic model, k_2 represents the adsorption rate constant of the pseudo-second order kinetic equation, which describes how fast the adsorption is carried out under the corresponding conditions. Table 5 suggests that when the initial phosphorus concentration was 1.0 mg/L, the k_2 was 0.742, which was the maximum value. The k_2 decreased with the increase in the initial P concentration. The k_2 was only 0.301 under the condition that the initial P concentration was 0.3 mg/L. These results further demonstrate that under reaction conditions with a high P concentration, the adsorption capacity of the adsorbent increases, and the efficiency of phosphorus removal also increases remarkably.

3.5. Effect of RSPRC Distribution in Reactor on Phosphorus Removal

As mentioned in Table 6, phosphorus removal performance reached a low point under experimental condition a. The concentration of effluent P amounted to 0.34 mg/L and the phosphorus removal rate was merely 32%. Based on further analysis, the adsorbent was concentratedly distributed between adjacent diversion walls. Although the adsorbent can absorb phosphorus in the nearby wastewater with a higher efficiency, overall, P concentration of the wastewater is lower, and the solute concentration gradient affecting the diffusion rate is smaller. Therefore, the phosphorus in the wastewater is diffused with difficulty and it is unable to sufficiently contact the adsorbent. Ultimately, the adsorption capacity of RSPRC is not fully utilized [31,32]. Under the experimental condition d, the equal spacing of the adsorbent is evenly distributed between the adjacent diversion walls, which increases the contact area with the sewage. As a result, the phosphorus adsorbate in wastewater can be efficiently absorbed [33] and the role of the adsorbent can be completely utilized to increase the unit adsorption amount of the adsorbent. The effluent P concentration in experimental condition d was the lowest at only 0.08 mg/L and the phosphorus removal rate reached up to 84%. Ragheb observed that the removal of P increased with the increase in rpm to some extent and attributed it to the dispersal of the adsorbent particles in the aqueous solution, which leads to a reduced boundary mass transfer [34]. The dispersive distribution of the adsorbent increases the removal rate, whose mechanism is similar to that of the increase in rpm. In an attempt to achieve the best static adsorption phosphorus removal result in the concentric circular diversion wall adsorption reactor, the experimental condition d was selected as the adsorbent distribution in the subsequent experiments.

Table 6. Effect of experimental condition in reactor on phosphorus removal.

Experimental Condition	Initial Concentration (mg/L)	Effluent Concentration (mg/L)	Phosphorus Removal Rate (%)
a	0.50	0.34	32
b	0.50	0.28	44
c	0.50	0.18	64
d	0.50	0.08	84

3.6. Effect of RSPRC Dosage in Reactor on Phosphorus Removal

Figure 9 presents the experimental data on the effect of RSPRC dosage in the reactor on phosphorus removal. Interestingly, it was found that the P removal rate was almost the same as that of the former agitation experiment. However, Rastas and Hedstrom hold the view that the acquired absolute values for adsorption capacities from agitation experiments cannot be extrapolated to practical applications. They explained that the direct contact between the grains and solution differs and the agitation of the blast furnace slag may cause the destruction of the material, which may increase the sorption sites and thus the sorption capacity could be overestimated [35]. A possible explanation for the difference might be that the reactor realizes multi-stage series adsorption, which improves the removal rate and makes up for the effect of insufficient contact. With the increase in initial P concentration, the concentration of effluent phosphorus decreases when the dosage of adsorbent increases. If the initial P concentration is larger, the trend tends to be more obvious [30,36]. Further analysis indicates that when the P concentration in the wastewater is reduced to a certain extent, that the phosphorus removal performance of the adsorbent in the wastewater begins to decrease. On the condition that the P concentration drops to a very low value (0.36 mg/L), the effect of the adsorbent in this study is greatly inhibited, and it is impossible to effectively reduce the P concentration in the wastewater.

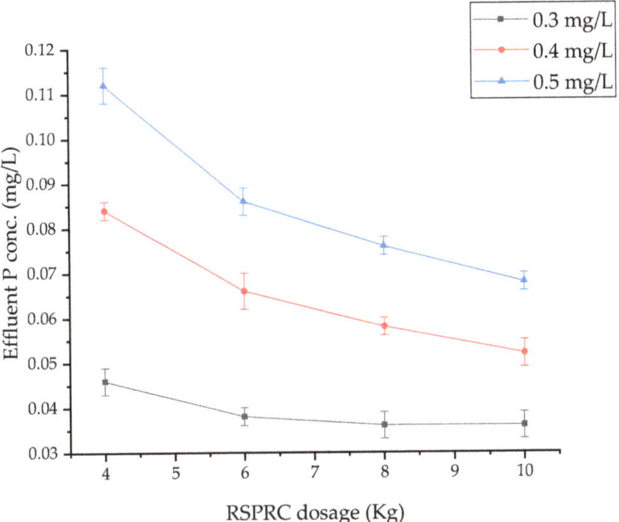

Figure 9. Effect of RSPRC dosage in the reactor on the effluent phosphorus concentration.

3.7. Effect of HRT in Reactor on Phosphorus Removal

The results of the effect of hydraulic retention time in reactor on phosphorus removal are presented in Figure 10. In cases where the HRT increases, the phosphorus removal rate increases significantly at a low adsorbent dosage, but with an increasing adsorbent dosage, the increase in the phosphorus removal rate of the adsorption reactor slows down [26,37]. The trend is similar to that of agitation experiments. Lu et al. achieved the same result. They explained that the rapid initial removal was mainly attributed to the precipitation of phosphate with exchangeable and dissolved Ca^{2+} rather than the adsorption. The liberated Ca^{2+} from the exchange site or from the dissolution of $CaCO_3$, CaO and $Ca(OH)_2$ were preferably precipitated by phosphate, which gave a high initial rate of adsorption. The exact contribution of the adsorption and precipitation phases to the removal of phosphate remains unclear [38]. Another possible explanation might be that when the HRT is short, the dosage of the adsorbent selected in the experiment cannot balance the adsorption reaction in the wastewater. In other words, the P concentration differs greatly from the equilibrium state. Although the reaction effect of the adsorbent has declined, the removal efficiency has not been excessively inhibited. An earlier study revealed that when the HRT is long, the instantaneous P concentration of the wastewater approaches the equilibrium state of the reaction, where the function of the adsorbent is obviously suppressed [39,40]. In cases where the dosage increases, the removal rate of phosphorus in the wastewater is still extremely slow. This is reflected in the fact that the longer the hydraulic retention time is, the smaller the influence of the adsorbent dosage on the P concentration of the effluent is.

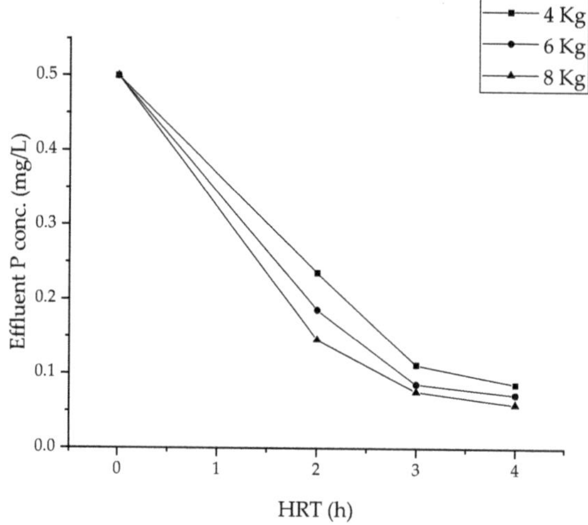

Figure 10. Effect of HRT in the reactor on the effluent phosphorus concentration.

4. Practical Applications and Future Research Perspective

It is economically feasible to use RSPRC as the adsorbent for the deep dephosphorization of municipal wastewater. The self-developed concentric circular diversion wall adsorption reactor can effectively and stably control the effluent phosphorus concentration, within 0.10 mg/L, and create no secondary pollution. In view of the deficiencies found in the research, combined with the practical needs of sewage treatment, these findings provide the following insights for future research. Firstly, further studies need to be carried out in order to modify the materials of the adsorbent to enhance the phosphorus removal performance. Secondly, it is necessary to optimize the material formulation, improve the manufacturing process, and increase the strength of RSPRC. Thirdly, the study should be repeated using different adsorbents in the self-developed concentric circular diversion wall adsorption reactor, which would be a fruitful area for further work. Finally, further research could usefully explore how to optimize the structure of the reactor so that the reactor can be flexibly changed in practical applications as required.

5. Conclusions

This study set out to prepare and apply RSPRC to remove phosphorus in the self-developed concentric circular diversion wall adsorption reactor. The second aim was to investigate the relationship between phosphorus removal effect and several influencing factors. One of the more significant findings to emerge from this study is that the addition of activated clay can make the adsorbent possess a good decolorization performance and effectively improve the phosphorus removal performance of the adsorbent. The second major finding was that both the pseudo-first order and pseudo-second order kinetic models can better reflect the adsorption kinetics of adsorbents in different initial P concentrations, but the latter are more accurate. This study has also found that the phosphorus removal rate is the highest on the condition that the adsorbent is evenly distributed between the adjacent diversion walls. The research results provide a valuable reference for practical wastewater applications, and a feasible method for the application of the adsorbent in the advanced treatment of wastewater.

Author Contributions: Conceptualization, Y.L.; methodology, Y.L. and L.Z.; software, Y.L. and L.Z.; validation, Y.L. and L.Z.; formal analysis, L.Z.; investigation, L.Z.; resources, Y.L.; data curation, L.Z.; writing—original draft preparation, L.Z. and R.P.S.; writing—review and editing, R.P.S. and L.Z.; visualization, Y.L. and L.Z.; supervision, Y.L. and R.P.S.; project administration, Y.L. All authors have read and agreed to the published version of the manuscript.

Funding: This research received no external funding.

Acknowledgments: The authors would like to show their gratitude to all those who helped during the experimental period and writing of this manuscript. L.Z. would like to gratefully acknowledge the help of Rajendra Prasad Singh, School of Civil Engineering, Southeast University. I appreciate his patience, encouragement and professional instructions.

Conflicts of Interest: The authors declare no conflict of interest.

References

1. Schindler, D.W.; Carpenter, S.R.; Chapra, S.C.; Hecky, R.E.; Orihel, D.M. Reducing phosphorus to curb lake eutrophication is a success. *Environ. Sci. Technol.* **2016**, *50*, 8923–8929. [CrossRef] [PubMed]
2. Mishra, S.; Kumar, A. Estimation of physicochemical characteristics and associated metal contamination risk in the Narmada River, India. *Environ. Eng. Res.* **2020**, *26*, 190521. [CrossRef]
3. Loganathan, P.; Vigneswaran, S.; Kandasamy, J.; Bolan, N.S. Removal and recovery of phosphate from water using sorption. *Crit. Rev. Environ. Sci. Technol.* **2014**, *44*, 847–907. [CrossRef]
4. Kumar, A.; Sharma, M.P.; Taxak, A.K. Analysis of water environment changing trend in Bhagirathi tributary of Ganges in India. *Desalin. Water Treat.* **2017**, *63*, 55–62. [CrossRef]
5. Kumar, A.; Sharma, M.P.; Rai, S.P. A novel approach for river health assessment of Chambal using fuzzy modeling, India. *Desalin. Water Treat.* **2017**, *58*, 72–79. [CrossRef]
6. Li, Y.; He, X.; Hu, H. Enhanced phosphate removal from wastewater by using in situ generated fresh trivalent Fe composition through the interaction of Fe(II) on CaCO3. *J. Environ. Manag.* **2018**, *221*, 38–44. [CrossRef]
7. Benammar, L.; Menasria, T.; Ayachi, A.; Benounis, M. Phosphate removal using aerobic bacterial consortium and pure cultures isolated from activated sludge. *Process Saf. Environ.* **2015**, *95*, 237–246. [CrossRef]
8. Wang, Y.; Liu, L.; Fan, W.; Yuan, L.; Luo, D. The mechanism of biological phosphorus removal under anoxic-aerobic alternation condition with starch as sole carbon source and its biochemical pathway. *Biochem. Eng. J.* **2018**, *132*, 90–99. [CrossRef]
9. Mitrogiannis, D.; Psychoyou, M.; Baziotis, I.; Inglezakis, V.J.; Koukouzas, N.; Tsoukalas, N.; Palles, D.; Kamitsos, E.; Oikonomou, G.; Markou, G. Removal of phosphate from aqueous solutions by adsorption onto Ca(OH)(2) treated natural clinoptilolite. *Chem. Eng. J.* **2017**, *320*, 510–522. [CrossRef]
10. Kuroki, V.; Bosco, G.E.; Fadini, P.S.; Mozeto, A.A.; Cestari, A.R.; Carvalho, W.A. Use of a La(III)-modified bentonite for effective phosphate removal from aqueous media. *J. Hazard. Mater.* **2014**, *274*, 124–131. [CrossRef]
11. Mulkerrins, D.; Dobson, A.; Colleran, E. Parameters affecting biological phosphate removal from wastewaters. *Environ. Int.* **2004**, *30*, 249–259. [CrossRef]
12. Lei, G.; Zhang, B.H. Progress of phosphorus wastewater treatment technology. *Energy Conserv. Emiss. Reduct. Pet. Petrochem. Ind.* **2013**, *3*, 37–40. (In Chinese)
13. Bhatnagar, A.; Sillanpaa, M. A review of emerging adsorbents for nitrate removal from water. *Biochem. Eng. J.* **2011**, *168*, 493–504. [CrossRef]
14. Ahmed, S.; Ashiq, M.N.; Li, D.; Tang, P.; Leroux, F.; Feng, Y. Recent progress on adsorption materials for phosphate removal. *Recent Pat. Nanotechnol.* **2019**, *13*, 3–16. [CrossRef]
15. Lakshmi, V.; Resmi, V.G.; Raju, A.; Deepa, J.P.; Rajan, T.P.D.; Pavithran, C.; Pai, B.C. Concentration dependent pore morphological tuning of kaolin clay foams using sodium dodecyl sulfate as foaming agent. *Ceram. Int.* **2015**, *41*, 14263–14269. [CrossRef]
16. Zhang, W. Progress in the application of bentonite to the adsorption for inorganic pollutants in water pollution control. *Ind. Water Treat.* **2018**, *38*, 10–16. (In Chinese)
17. Wang, T.; Liu, T.; Wu, D.; Li, M.; Chen, J.; Teng, S. Performance of phosphoric acid activated montmorillonite as buffer materials for radioactive waste repository. *J. Hazard. Mater.* **2010**, *173*, 335–342. [CrossRef]
18. Wu, L.L.; Liu, Y.; Wang, S.H.; Zhang, H.; Huang, M. Effects of the particle size and pH value on EPRC in phosphate removal. *Saf. Environ. Eng.* **2008**, *15*, 45–48. (In Chinese)
19. Zhou, J. Study on Phosphorus Removal Characteristics of Perforated EPRC. Master's Thesis, Southeast University, Nanjing, China, 2010. (In Chinese)
20. *Standard Methods for the Examination of Water and Wastewater*; APHA (American Public Health Association): Washington, DC, USA, 1989.

21. Gamba, M.; Kovar, P.; Pospisil, M.; Torres Sanchez, R.M. Insight into thiabendazole interaction with montmorillonite and organically modified montmorillonites. *Appl. Clay Sci.* **2017**, *137*, 59–68. [CrossRef]

22. Ashekuzzaman, S.M.; Jiang, J. Study on the sorption-desorption-regeneration performance of Ca-, Mg- and CaMg-based layered double hydroxides for removing phosphate from water. *Biochem. Eng. J.* **2014**, *246*, 97–105. [CrossRef]

23. Patra, B.S.; Baliarsingh, N.; Parida, K.M.; Das, J. Adsorption of phosphate by layered double hydroxides in aqueous solutions. *Appl. Clay Sci.* **2006**, *32*, 252–260.

24. Wang, H.Z.; Song, W.D.; Xing, K.; Zhao, Z.P.; Guo, L.G. Adsorption of tripolyphosphate from aqueous solution by Mg-Al-CO3-layered double hydroxides. *Colloids Surf. A* **2008**, *328*, 15–20.

25. Li, J.; Wu, B.; Zhou, T.; Chai, X. Preferential removal of phosphorus using modified steel slag and cement combination for its implications in engineering applications. *Environ. Technol. Innov.* **2018**, *10*, 264–274. [CrossRef]

26. Shan, H.D.; Lu, S.G.; Bai, S.Q.; Zhu, L. Removal mechanism of phosphate from aqueous solution by fly ash. *J. Hazard. Mater.* **2009**, *161*, 95–101.

27. Agyei, N.M.; Strydom, C.A.; Potgieter, J.H. The removal of phosphate ions from aqueous solution by fly ash, slag, ordinary Portland cement and related blends. *Cem. Concr. Res.* **2002**, *32*, 1889–1897. [CrossRef]

28. Li, X.; Wang, G.; Li, W.; Wang, P.; Su, C. Adsorption of acid and basic dyes by sludge-based activated carbon: Isotherm and kinetic studies. *J. Central South. Univ.* **2015**, *22*, 103–113. [CrossRef]

29. Zhao, Y.Q.; Babatunde, A.O. Equilibrium and kinetic analysis of phosphorus adsorption from aqueous solution using waste alum sludge. *J. Hazard. Mater.* **2010**, *184*, 746–752.

30. Meyer, D.; Andres, Y.; Chazarenc, F.; Barca, C.; Gerente, C. Phosphate removal from synthetic and real wastewater using steel slags produced in Europe. *Water Res.* **2012**, *46*, 2376–2384.

31. Liu, Y.; Li, L.; Zhang, S.J.; Wang, S.H. Kinetics and thermodynamics of efficient phosphorus removal by a complex material. *Desalin. Water Treat.* **2015**, *56*, 1949–1954. [CrossRef]

32. Yao, X.D.; Haghseresht, F.; Wang, S.B.; Rudolph, V.; Zhu, Z.H.; Li, L.; Huang, W.W. Phosphate removal from wastewater using red mud. *J. Hazard. Mater.* **2008**, *158*, 35–42.

33. Goscianska, J.; Ptaszkowska-Koniarz, M.; Frankowski, M.; Franus, M.; Panek, R.; Franus, W. Removal of phosphate from water by lanthanum-modified zeolites obtained from fly ash. *J. Colloid Inter. Sci.* **2018**, *513*, 72–81. [CrossRef] [PubMed]

34. Ragheb, S.M. Phosphate removal from aqueous solution using slag and fly ash. *HBRC J.* **2013**, *9*, 270–275. [CrossRef]

35. Rastas, L.; Hedstrom, A. Methodological aspects of using blast furnace slag for wastewater phosphorus removal. *J. Environ. Eng.* **2006**, *132*, 1431–1438.

36. Molle, P.; Saenz De Miera, L.E.; Blanco, I.; Ansola, G. Basic Oxygen Furnace steel slag aggregates for phosphorus treatment. Evaluation of its potential use as a substrate in constructed wetlands. *Water Res.* **2016**, *89*, 355–365.

37. Senan, P.; Anirudhan, T.S. Adsorption of phosphate ions from water using a novel cellulose-based adsorbent. *Chem. Ecol.* **2011**, *27*, 147–164.

38. Lu, S.G.; Shan, H.D.; Bai, S.Q. Mechanisms of phosphate removal from aqueous solution by blast furnace slag and steel furnace slag. *J. Zhejiang Univ. Sci. A* **2008**, *9*, 125–132. [CrossRef]

39. Pengthamkeerati, P.; Satapanajaru, T.; Chularuengoaksorn, P. Chemical modification of coal fly ash for the removal of phosphate from aqueous solution. *Fuel* **2008**, *87*, 2469–2476. [CrossRef]

40. Kumar, A.; Mishra, S. Environmental quantification of soil elements in the catchment of hydroelectric reservoirs in India. *Hum. Ecolog. Risk Asses* **2017**, *23*, 1202–1218. [CrossRef]

MDPI
St. Alban-Anlage 66
4052 Basel
Switzerland
Tel. +41 61 683 77 34
Fax +41 61 302 89 18
www.mdpi.com

Applied Sciences Editorial Office
E-mail: applsci@mdpi.com
www.mdpi.com/journal/applsci